KT-416-884

GLOBAL WARMING

FOR BEGINNERS®

C0000 002 421 403

HOUNSLOW LIBRARIES	
BED	
C0000 002 421 403	
Askews	18-Mar-2009
363.7387	£9.99

GLOBAL WARMING
FOR BEGINNERS®

BY **DEAN GOODWIN, Ph.D.**
ILLUSTRATIONS BY **JOE LEE**

FOR BEGINNERS®

an imprint of Steerforth Press

Hanover, New Hampshire

For Beginners LLC
62 East Starrs Plain Road
Danbury, CT 06810 USA
www.forbeginnersbooks.com

Text: © 2008 Dean Goodwin, Ph.D.
Illustrations: © 2008 Joe Lee
Book Design: David Janik

This book is sold subject to the condition that it shall not, by way of
trade or otherwise, be lent, re-sold, hired out, or otherwise circulated
without the publisher's prior consent in any form of binding or cover other
than that in which it is published and without a similar condition being
imposed on the subsequent purchaser.

All rights reserved. No part of this publication may be reproduced, stored
in a retrieval system, or transmitted in any form or by any means,
electronic, mechanical, photocopying, recording, or otherwise, without prior
permission of the publisher.

A For Beginners® Documentary Comic Book
Copyright © 2008

Cataloging-in-Publication information is available from the Library of Congress.

ISBN # 978-1-934389-27-0 Trade

Manufactured in the United States of America

For Beginners® and Beginners Documentary Comic Books® are published
by For Beginners LLC.

First Edition

10 9 8 7 6 5 4 3 2 1

Back Cover: Environmental impact estimates were made
using the Environmental Defense Fund Paper Calculator.
For more information visit http://www.papercalculator.org.

Printed on 100% postconsumer recycled content paper.

Contents

Chapter 1

Global Warming: An Introduction

It's Such a Hot Topic!

Global warming is a major environmental issue, one that has generated much interest, debate, argument, and conjecture in the media in recent times. Twenty years ago the general public had little idea of its existence; today the topic receives daily media attention. Some scientists have stood by the evidence that proves global warming while others have got very hot under the collar in their attempts to convince us otherwise. But the debate has not been just among scientists (where it is to be expected, because the scientific process should be open to healthy skepticism and data should be able to stand up to scrutiny by the whole global scientific community). Others who may not be as knowledgeable about the subject have gotten into the discussion, letting off enough steam that their voices are heard. This has led to the spread of misinformation and misconceptions in the eyes of the public as the global warming and climate change debate spirals out of control, leading to a lot of confusion and misunderstanding.

So Why All the Fuss?

Many scientists have suggested that the climate changes that are occurring as a result of global warming are being caused by us—the human race. Societies around the world are using more and more fossil fuels, such as coal, oil, and natural gas, to provide increasing amounts of energy in order to meet the demands of an increasing population while at the same time increasing economic growth. Whenever a fossil fuel is burned it produces, among other things, carbon dioxide, which gets released into the atmosphere. Carbon dioxide and some other compounds have been shown to be capable of raising the temperature of the atmosphere and have been labeled *greenhouse gases*. Many scientists are calling for a reduction in the amounts of carbon dioxide and the other greenhouse gases that we are producing in order to slow down the rate of global warming and the accompanying changes in the earth's climate. Reducing carbon dioxide emissions would necessitate a reduction in the use of fossil fuels or a shift to using them more efficiently. That's why the debate really began.

Many untruths and junk science are dispersed to the public through the media by special interest groups that are directly involved in the production or use of fossil fuels, including energy utility companies. These groups rationalized that if carbon dioxide has to be reduced, then new regulations and mandates will adversely affect their operations. They disagree with the idea of global warming and the role that carbon dioxide plays in the process. The misinformation put forward by such groups not only receives widespread attention in the media, it also has the support of some politicians and other scientists, many of whom are not climatologists. They argue that climate change is part of the earth's natural cycles, and that the increase in amounts of greenhouse gases in the atmosphere during the past 200 years or so are not to blame for the increased planetary temperatures.

> "Global warming is the second-largest hoax played on the
> American people, after the separation of church and state."
> —US Senator James M. Inhofe (R-Oklahoma), 2005

Putting Things into Perspective

It is true that over many thousands of years the earth experiences dramatic climate changes, and there have been some hot and cold anomalies that do occur once in a while. For example, the folks in New England called 1816 "the year without a summer," because snow fell in every month of that year! In 17th-century Europe the temperatures were so low that England's River Thames froze, and the term "Little Ice Age" was used. Yes, these anomalies could have resulted from causes that are unrelated to the carbon dioxide levels in the atmosphere, such as decreased sunspot activity, just as abnormally warm periods could occur due to increased sunspot activity. So today the skeptics use such arguments to discount the current climate changes as being just a part of the Earth's natural cycles and nothing for us to worry about. Not only that, they claim we are not responsible for these changes, and what's more, carbon dioxide is a gas that occurs naturally so how can our emissions of this natural substance affect anything? These arguments can sound very convincing, especially if one has little understanding of or basic common grounding in science.

However, people were worrying about humans' affects on nature long before the global warming debate began.

In an address delivered before the Agricultural Society of Rutland County, Vermont, on September 30, 1847, George Perkins Marsh said, "But though man cannot at his pleasure command the rain and the sunshine, the wind and frost and snow . . . it is certain that climate itself has in many instances been gradually changed and ameliorated or deteriorated by human action." Marsh was a farmer, and probably the first person to refer to human-induced climate change. He wrote a physical geography book in 1864, *Man and Nature*, in which he said the following:

GEORGE PERKINS MARSH (1801-1882)

The action of man, indeed, is frequently followed by unforeseen and undesired results, yet it is nevertheless guided by a self-conscious will . . . .
The ravages committed by man . . . destroys the balance which nature had established . . . and she avenges herself . . . by letting loose . . . destructive energies hitherto kept in check . . . but which he has unwisely dispersed.
The earth is fast becoming an unfit home for its noblest inhabitant, and another era of equal human crime . . . would reduce it to such a condition of impoverished productiveness, of shattered surface, of climate excess, as to threaten the depravation, barbarism, and perhaps even extinction of the species.

From these comments, it would appear that Marsh could be considered the Nostradamus of climate change and global warming, with his prophesies gaining support from the scientific community some 150 years later! Although he did not link the role of carbon dioxide to global warming, he did make a strong case for how changes in land use and human activity can adversely affect climate.

So How Did Carbon Dioxide Get the Blame?

Over the years many scientists have studied the properties of carbon dioxide. In 1860, John Tyndall, a British physicist born in Ireland, conducted experiments to measure the amount of infrared radiation (heat) that carbon dioxide could absorb.

JOHN TYNDALL
(1820-1893)
SUPERINTENDENT:
THE ROYAL INSTITUTION
OF GREAT BRITAIN

Earlier, in 1827, Jean Baptiste Joseph Fourier, a French mathematician who researched heat conduction, had suggested that heat gets trapped near the earth because the atmosphere behaves like "the glass of a hothouse." This paved the way for the now ubiquitous phrase "greenhouse effect."

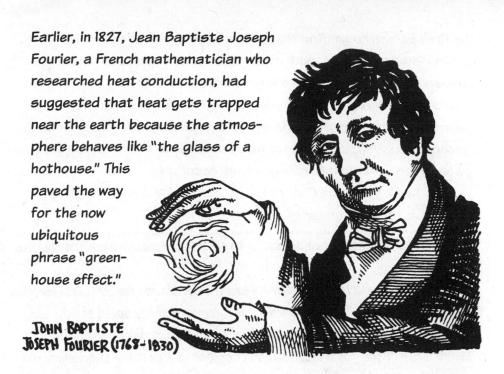

JOHN BAPTISTE JOSEPH FOURIER (1768-1830)

In 1894, Swedish physical chemist Svante Arrhenius hypothesized that the increased emission of carbon dioxide as a result of the industrial revolution would result in global warming. He published the first calculations of increased atmospheric temperatures resulting from human-induced, or *anthropogenic*, increases in atmospheric carbon dioxide, and made predictions as to the extent of future temperature changes.

SVANTE ARRHENIUS
(1859 - 1927)

The first person to confirm that carbon dioxide released through the burning of fossil fuels and other industrial processes increased atmospheric levels of the gas was Charles David Keeling, who began recording the levels of carbon dioxide in the atmosphere above the Mauna Loa Observatory in Hawaii in the late 1950s. His landmark research laid the foundation for many studies of global carbon dioxide levels and the climate changes that occur as a result.

Upon Keeling's death, Charles F. Kennel, the director of the Scripps Institution for Oceanography said:

> There are three occasions when dedication to scientific measurements has changed all of science.
>
> Tycho Brahe's observations of planets laid the foundation for Sir Isaac Newton's theory of gravitation. Albert Michelson's measurements of the speed of light laid the foundation for Albert Einstein's theory of relativity. Charles David Keeling's measurements of the global accumulation of carbon dioxide in the atmosphere set the stage for today's profound concerns about climate change. They are the single most important environmental data set taken in the 20th century.

This data set is widely known today as the "Keeling Curve" and is referenced in many articles, textbooks, and other media presentations on global warming.

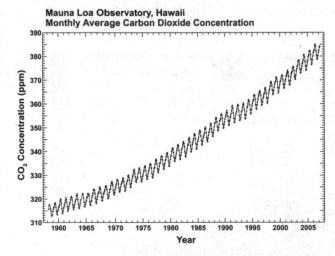

Comparing the Keeling Curve to other graphs showing the increase in atmospheric temperatures has enabled climatologists to clearly establish the link between greenhouse gas pollution and global warming.

The Keeling Curve, showing the increased atmospheric carbon dioxide levels over the last half-century.

CHARLES DAVID KEELING (1928-2005)

I HAVE THE MEASURE OF IT.

In the 1950s, climatologists began to use computer models to predict changes in the earth's climate as a result of global warming. In the 1980s it was clear that an actual warming trend was emerging. In 1988, the World Meteorological Organization (WMO) and the United Nations Environmental Programme (UNEP) established the Intergovernmental Panel on Climate Change (IPCC) to investigate and report back their findings on increasing global temperatures due to increased greenhouse gas emissions.

What Is the IPCC and What Has It Reported?

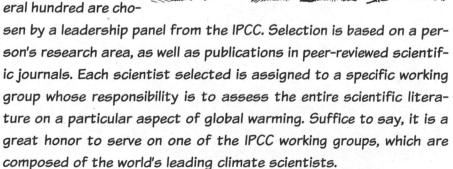

The IPCC is, basically, a group of scientists. Governments of different countries submit the names of their own climatologists, and then several hundred are chosen by a leadership panel from the IPCC. Selection is based on a person's research area, as well as publications in peer-reviewed scientific journals. Each scientist selected is assigned to a specific working group whose responsibility is to assess the entire scientific literature on a particular aspect of global warming. Suffice to say, it is a great honor to serve on one of the IPCC working groups, which are composed of the world's leading climate scientists.

9

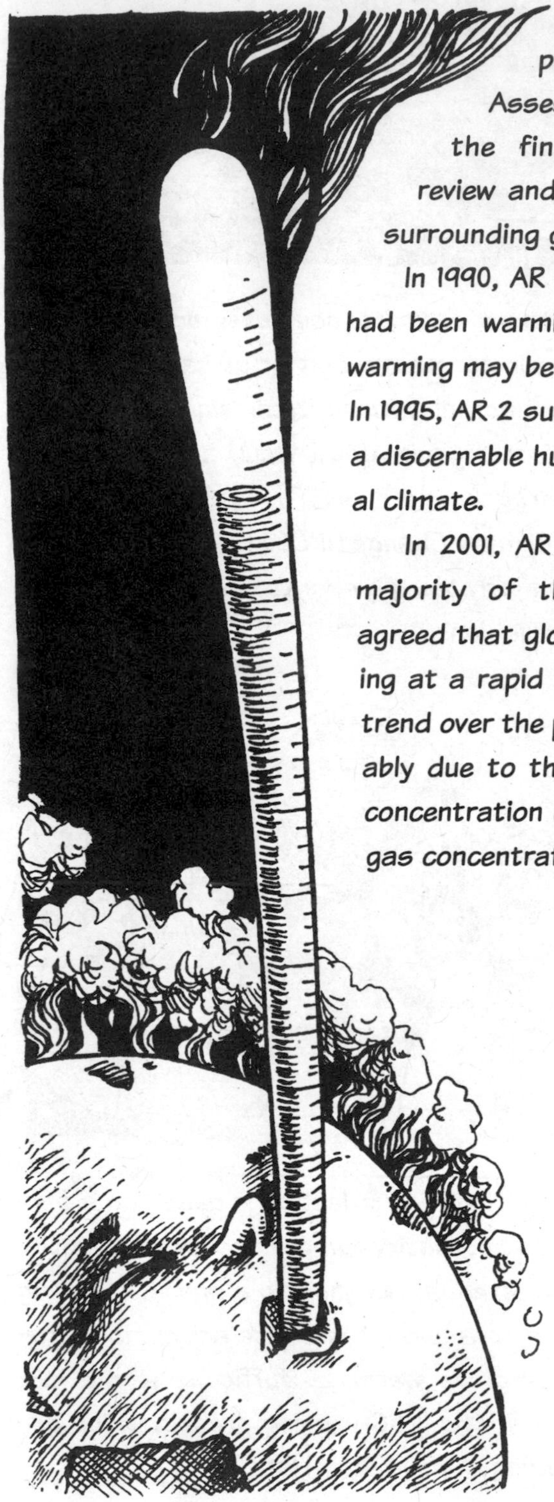

To date the IPCC has produced and published four Assessment Reports (ARs) on the findings of their intensive review and scrutiny of the literature surrounding global warming.

In 1990, AR 1 concluded that the earth had been warming and that a lot of the warming may be due to natural processes. In 1995, AR 2 suggested that there may be a discernable human influence on the global climate.

In 2001, AR 3 reported that the vast majority of the world's climatologists agreed that global temperatures were rising at a rapid rate and that the warming trend over the past half century was probably due to the observed increase in the concentration of atmospheric greenhouse gas concentrations. The IPCC concluded:

Since the late 1950s, the overall global temperatures in the lowest part of the atmosphere and on the earth's surface have been rising at a rate of 0.1 degree centigrade per decade.

The atmospheric concentration of carbon dioxide has increased 31% since 1750. The present carbon dioxide

concentration has not
been exceeded during the past
420,000 years and likely not during the past 20
million years. The current rate of increase is unprecedented during at least the past 20,000 years.

About three-quarters of the anthropogenic emissions of carbon dioxide to the atmosphere during the past 20 years
is due to fossil-fuel burning. The rest is predomi-
nantly due to land use change, especially
deforestation.

These findings caused
alarm and concern in
many quarters.
Environmental organizations
became increasingly at odds with the fossil
fuel industry and energy producing companies, who
firmly denied their role in climate change. Scientific organizations in the United States, including the National Academy of
Sciences and the American Association for the Advancement of
Science, began to adopt the view put forward by the IPCC. After
reviewing all of the data, eminent scientist Dan Agin, a longtime

promoter of scientific literacy, concluded that unless a group of mysterious aliens continually visiting us during the past 250 years spritzed carbon dioxide into our atmosphere, the only sensible conclusion from the above data is that the increase in atmospheric carbon dioxide concentration has been due to anthropogenic emissions, particularly those associated with the Industrial Revolution.

The evidence became even stronger with publication of the IPCC's AR 4 in November 2007. This report reflected the work of more than 2,500 expert reviewers, 800 contributing authors, and 450 lead authors from 130 countries. The report states:

> Warming of the climate system is unequivocal, as is now evident from observations of increases in global average air and ocean temperatures, widespread melting of snow and ice, and rising global average sea level.

> Observational evidence from all continents and most oceans shows that many natural systems are being affected by regional climate changes, particularly temperature increases.

> Global greenhouse gas emissions due to human activities have grown since pre-industrial times, with an increase of 70% between 1970 and 2004.

> Most of the observed increase in globally-averaged temperatures since the mid-20th century is very likely due to the observed increase in anthropogenic greenhouse gas concentrations.

> Anthropogenic warming over the past three decades has likely had a discernable influence at the global scale on observed changes in many physical and biological systems.

Prior to the announcement of the AR 4 findings by Chairman Rajendra Pachauri at the 27th session of the IPCC in Valencia, Spain, UN Secretary-General Ban Ki-moon introduced the report by saying:

> Climate change is a serious threat to development every-
> where. Today, the time for doubt has passed. The IPCC
> has unequivocally affirmed the warming of our climate
> system, and linked it directly to human activity. Slowing or
> even reversing the existing trends of global warming is the
> defining challenge of our ages.

On December 10, 2007, in Oslo, Norway, the IPCC and Albert Gore, Jr., were awarded the Nobel Peace Prize "for their efforts to build up and disseminate greater knowledge about man-made climate change, and to lay the foundations for the measures that are needed to counteract such change."

So Where Do We Go from Here?

The time for debate about the role of carbon dioxide and other greenhouse gases in climate change and global warming is over. The scientific consensus has firmly established that global warming is occurring and that human activities play a major role in causing it. We know that the evidence is real, and we need to accept and believe the majority of the world's scientists. The public should no longer accept messages about global warming that are flawed by untruths and misconceptions, promulgated by people or special interest groups who do not fully understand or want to accept the scientific explanation of this observed fact: Global carbon dioxide levels are increasing due to human activities, and are causing the earth's temperature to rise. Just as scientists have shown that the earth is not flat, that the sun does not revolve around the earth, and that humans were not around when the dinosaurs roamed some 65 million years ago (despite what you see in the movies), we must also accept the reality of global warming.

"The skeptics have had their heyday. This is abundantly clear. Nobody's questioning the science anymore."
—Yvo de Boer, Executive Secretary of the United Nations Framework Convention on Climate Change, Bali, Indonesia, December 2007

Since the question of global warming and its causes are no longer a subject for debate, the issue now is what to do to address the problem. The question is no longer "Is this happening?" The question now is "What can we do about it?" When Rajendra Pachauri presented the conclusions from AR 4 he posed some more specific questions:

> How do we prepare the human race to face sea level rise and a world with new geographical features? Is the current pace and pattern of development sustainable? What changes in lifestyles, behavior patterns, and management practices are needed, and by when?

The answers will ultimately depend upon the policies developed by the governments around the globe. How well these governments represent their citizens in addressing global warming remains to be seen. It is incumbent upon every citizen in each country to fully understand the science behind global warming, its causes and consequences, and the role we all play in working toward a solution. This is where the road ahead will be hard traveling, and there will be much debate, discussion, and argument.

The first step down that road is to educate yourself and learn as much as possible about global warming. The chapters that follow will help you begin a journey toward scientific literacy on the subject, engender understanding, and provide the tools you'll need to take action as a public citizen. Remember: The solution to

"WE HAVE NOT INHERITED THE WORLD FROM OUR ANCESTORS; WE HAVE BORROWED IT FROM OUR CHILDREN."
NATIVE AMERICAN QUOTE

global warming depends ultimately on concerned, well-informed citizens. As knowledgeable citizens we have the power to elect officials who will best represent our views, who in turn determine the government policies that will affect everyone on the planet now and in the future. We owe that to our children and those who follow after them.

As a well-informed citizen, you will be able to take your place at the table alongside the scientists, politicians, and leaders from business and industry as solutions to global warming are debated. You may even be able to convince the politicians, environmentalists, and industrialists that they should no longer see themselves as being on opposite sides of the table. In fact, we are all at the same table and it is round—we are in this together and we must all work together, accepting compromises. In the words of Mahatma Ghandi, "Be the change you want to see in the world."

Change Is in the Air!

All over the world, individuals, governments, and business and industry leaders have already begun to move toward global warming solutions. They use energy more efficiently and make increased use of renewable and alternative energy sources. This not only makes economic sense, because it saves money, but also helps the environment in the process. We must be creative in how we solve the problems associated with climate change and global warming, using critical thinking, teamwork, compromise through effective communication, and global cooperation. As Albert Einstein reminds us, "The significant problems we face cannot be solved with the same level of thinking we used when we created them."

Examples of such steps forward are easy to find: Australia has decided to ban the use of incandescent light bulbs by 2009. Canada is ready to follow suit in 2012. The European Union plans to increase the use of renewable energy to 12 percent by 2010. Through improving energy efficiency Germany plans to cut its greenhouse gas emissions 65 percent by 2050. Government-owned power companies in China are building inland and coastal wind farms. China has also introduced stricter fuel-efficiency standards.

In the United States, an organization comprised of more than thirty leading businesses and environmental groups has recently been established. The US Climate Action Partnership (USCAP) is "committed to a pathway that will slow, stop, and reverse the growth of US emissions while expanding the US economy." They go on to say "in our view, the climate change challenge will create more economic opportunities than risks for the US economy." One of the group's missions is to urge the federal government to act quickly and "enact strong national legislation to require significant reductions of greenhouse gas emissions."

Late in the evening on December 13, 2007, the US Senate approved, through bipartisan support, an energy bill that is a small first step in fulfilling USCAP goals and reducing greenhouse gas emissions. Some of the bill's provisions include raising the fuel efficiency standards for new automobile fleets to 35 miles per gallon by 2020, expanding the use of ethanol and other biofuels, and increasing the energy efficiency standards for household appliances and buildings.

This is the first time since 1975 that Congress has increased the fuel efficiency standards. Senate Majority Leader Harry Reid (D-Nevada) said the bill "will save consumers money, it will begin to reverse our addiction to oil, and it takes a small first step in our fight to turn the tide of global warming."

Senate Minority Leader Mitch McConnell (R-Kentucky) added "the new fuel economy standards and the increase in renewable fuels represent a step forward in our common effort to make America more energy independent." The bill is not without critics, who feel that it could have been stronger in terms of fuel efficiency standards and are concerned that the increased use of corn to produce ethanol could have detrimental environmental effects, such as increased water use and fertilizer runoff problems, in addition to raising food prices and costing billions of dollars in federal subsidies. It was hoped that the bill would require utilities to generate 15 percent of their power from renewable energy, but this

measure was
dropped. Gregory Wetstone
of the American Wind Energy
Association commented that "for the
wind industry, it looks like coal in our
Christmas stocking." Daniel Weiss of the
Center for American Progress agreed, saying the Senate had given "the green light
to more energy efficient cars and renewable fuels but has a red light for renewable energy from wind, solar, and
other clean sources." But at
least in the eyes of climate policy analyst for the Natural
Resources Defense Council,
Elizabeth Martin-Perera,
"Congress should be congratu-
lated for taking the first step in the
sprint to solve global warming."

The debate about how we can work
together to address the problem of global
warming is only just beginning. It will be an
interesting topic to follow over the coming
years, with twists and turns taking place on an
almost daily basis.

A major turnaround in the US policy toward
global warming came at the United Nations
Framework Convention on Climate Change in
Bali, Indonesia, on December 14, 2007. After
many emotive discussions, the US delegation agreed that it would
support the revised text of the "Bali road map." This paves the way for
a two-year process of meetings, talks, and negotiations designed to

produce a new set of global emission targets, replacing those stated in the Kyoto Protocol. At the end of the conference, Indonesian Environment Minister Rachmat Witoelar commented, "This is a real breakthrough, a real opportunity for the international community to successfully fight climate change." The US decision to support the next steps in tackling the global warming issue in such a public forum, after initially voting against them in Kyoto and Rio De Janeiro, marked a huge change in policy. At one point during the Bali conference one of Papua New Guinea's delegation members told the US delegation, "If you are not willing to lead, please get out of the way." Soon afterward the US finally voted in approval of the document.

The Bali road map addresses such topics as cutting greenhouse gas emissions, developing clean technologies (particularly in the developing world), bringing a cessation to deforestation, and assisting poor countries by protecting their economies and people against the consequences of the climate change, such as lower crop yields and higher sea levels. The British Prime Minister Gordon Brown said, "This is a vital step forward for the whole world. The Bali road map agreed [to] today is just the first step. Now begins the hardest work, as all nations work towards a deal in Copenhagen in 2009 to address the defining challenge of our time."

It is very easy to become overwhelmed: What role can any one individual have in working toward solving global warming? It is easier to give up hope and think that the problem will go away. But remember, it will take all of us working together as a team to elicit change. Environmentalists and business leaders, private citizens and politicians, the governments of every country in the world need to rise to the challenges that lie ahead for all of us. No one individual can do everything, but every individual can do something.

" THE BIGGEST THREAT TO HUMANITY IS NOT THE EVIL OF BAD PERSONS; IT'S THE PASSIVITY OF GOOD ONES."
MARTIN LUTHER KING JR.

Global Warming: The Cause

Past Climate Change

The earth's climate has changed greatly over millions of years. Long periods of global cooling resulted in ice ages and glaciers covered the land. Long periods of global warming resulted in thawing ice sheets and higher sea levels. What causes the natural fluctuation in the earth's climate?

Many things can lead to climate variations. Volcanic eruptions release particles into the lower atmosphere, or *troposphere*, which causes a cooling effect. Changes in the output of energy from the sun during sunspot activity can lead to temperature variations on Earth. The continents move slowly over long periods of time as the earth's tectonic plates adjust, which leads to alteration in the climate of the continents. The planet has been hit by large meteors, which leave behind large craters, generating a lot of dust and other particulate matter.

The earth's position in space can also lead to changes in climate. The earth is tilted on its axis at an angle of 23.5 degrees, the reason behind the seasonal changes in the Northern and Southern hemispheres. Every 41 thousand

years or so a shift in the earth's tilt changes the intensity of the sunlight that reaches the earth, which may start an ice age. The shape of the earth's orbit changes from more elliptical to more circular every hundred thousand years, which also leads to differences in the amount of solar energy reaching the planet. The earth wobbles on it axis just like a spinning top. Every 20,000 years or so it shifts back and forth, bringing about change in the seasonal climate in both hemispheres. These natural climate fluctuations have been used by skeptics to suggest that we should not worry about human-produced carbon dioxide affecting the earth's temperature, because any or all of these natural causes may be the culprit.

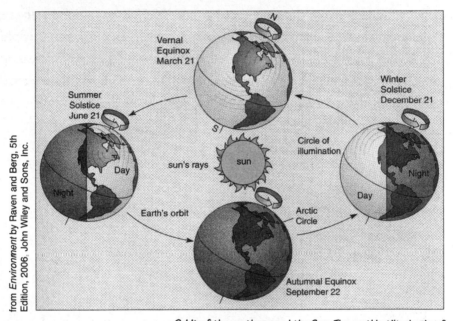

from *Environment* by Raven and Berg, 5th Edition, 2006, John Wiley and Sons, Inc.

Orbit of the earth around the Sun: The earth's tilted axis of 23.5 degrees creates seasonal changes in both hemispheres

How do we know that the climate can be affected by the atmospheric carbon dioxide level, in the past and in more recent times? Historic carbon dioxide levels can be accurately measured by a number of techniques. Some climatologists measure radioisotopes in rocks, fossils, and ocean sediments that contain coral and plankton. The thickness of tree rings can indicate faster growth, and maybe warmer years. Written records from monasteries and the like are also troves of information about historical temperatures. Ancient air, trapped in glaciers and ice sheets can be sampled from ice cores. Current changes in the earth's surface, ocean, and air temperatures have been monitored by weather balloon and satellite.

What can we find in these measurements? Analysis of ice core samples from the Antarctic have shown that over the past 400,000 years there is a high degree of correlation between the rise and fall of carbon dioxide levels and the changes in global temperature and sea level: more carbon dioxide, warmer temperatures and higher sea levels. The Antarctic data also show that long ice ages are followed by shorter warming periods. The warm periods, or interglacials, last about 10,000 years and occur about every 100,000 years. When scientists look at data from the most recent centuries, they see a rapid rise in carbon dioxide levels and temperatures rising at an increasing rate. In fact, combining ice core data with current measurements show that the tropospheric levels of carbon dioxide in 2005 were higher than at any point in the last 650,000 years. Climatologists at the NASA Goddard Institute for Space Studies found 2005 to be the warmest year on record since the 1890s.

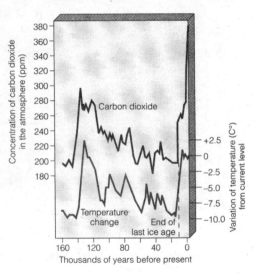

Correlation between tropospheric carbon dioxide level and global temperature near the earth's surface.

from *Living in the Environment* by Miller, 15th Edition, 2007, Thomson Higher Education.

The Greenhouse Effect

In order to understand the greenhouse effect it is important to first understand what can happen to solar radiation as it enters the earth's atmosphere.

The atmosphere is made up of several layers. If you think of the earth as a beach ball, the atmosphere would only be as thick as a paper towel! If the Earth were an apple, the atmosphere would be no thicker than its skin. Although the atmosphere seems a small player, it is an important part of the biosphere, responsible for the existence of life on Earth.

The troposphere is the layer closest to the earth, extending to a height of about 11 miles above the equator. All of our weather takes place in this layer. The stratosphere lies above the troposphere and reaches a height of about 31 miles. This layer is important, because it contains a region where the concentration of ozone is high (the reason it is known as the *ozone layer*).

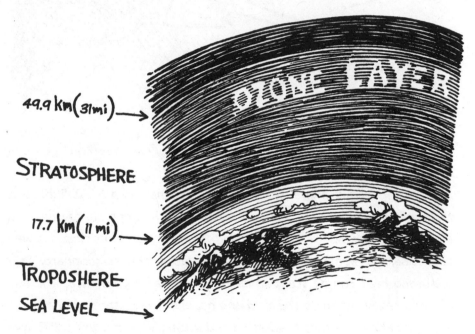

Lower layers of the atmosphere
As the solar radiation enters Earth's atmosphere a number of outcomes are possible. In the stratosphere some of the ultraviolet radiation is absorbed by the ozone layer. About 25 percent of the sunlight will be reflected back into space by clouds and particles in the atmosphere; some five percent will be reflected off surfaces such as ice and snow. Roughly 20 percent is absorbed directly by molecules in clouds and

particles in the atmosphere, and the remaining sunlight is absorbed by the land and oceans. It is this 50 percent that heats the earth.

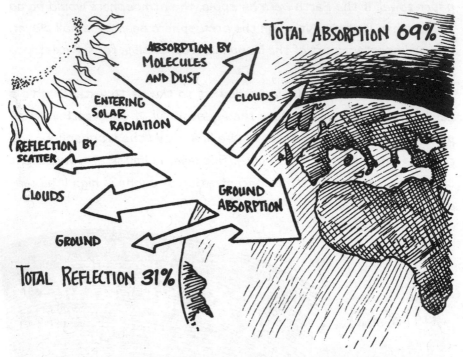

TOTAL ABSORPTION 69%

ABSORPTION BY MOLECULES AND DUST

ENTERING SOLAR RADIATION

CLOUDS

REFLECTION BY SCATTER

CLOUDS

GROUND ABSORPTION

GROUND

TOTAL REFLECTION 31%

The fate of solar energy reaching the earth

During daylight hours the land and oceans absorb sunlight and warm up. At night these warmed areas give the heat energy back to the atmosphere as infrared radiation. Some of this heat finds its way to space, but some is absorbed by gases in the atmosphere. This absorbed heat can be trapped close to the earth's surface, where it can once again warm up the land and oceans. The gases that have the ability to absorb infrared radiation and warm the planet are collectively termed *greenhouse gases*, because they act like the glass in a greenhouse, trapping heat. Think of your car, left out in the sun on a hot day with the windows closed. Sunlight passes through the glass and heats the interior of the car, causing the air temperature inside the car to increase. The infrared radiation (heat) emitted off the dashboard, seats, and so forth is reflected back into the car by the glass, causing the inside air to get hotter and hotter.

Greenhouse Gases

The atmosphere consists mostly of nitrogen (78 percent) and oxygen (21 percent), which are not very good at absorbing heat. Naturally occurring gases such as carbon dioxide, water vapor, methane, and nitrous oxide make up a small amount of the remaining one percent and are excellent at absorbing heat. These gases have been a part of the earth's atmosphere for millions of years, and over time their relative concentrations have fluctuated. This, together with variations in solar output, has caused the increase and decrease in the temperature of the atmosphere that resulted in past climate changes.

HEY, DON'T BLAME THIS ON US.

Water Vapor

Water is great at absorbing heat. Burn a finger on a hot saucepan and what do we do? Rush to put it under the cold tap. The water takes away some of the heat, providing relief from the pain. The coolant used in most combustion engines is water, because it can absorb a lot of heat from the engine without a large increase in the water temperature. The heat from the engine is passed out to the air by the water as it cycles through the radiator. Water vapor in the atmosphere can absorb a lot of heat, so too can the water in the oceans, lakes, rivers, and streams. It is hard to determine the actual extent of water's contribution to the recent increased warming of the planet, because it has always been present in the form of liquid (oceans), solid (ice and glaciers), and gas (water vapor). One thing is true: As the earth warms, more water will evaporate from the surface of lakes and oceans and become atmospheric water vapor.

Carbon Dioxide

Carbon dioxide is an odorless, tasteless, colorless gas that is constantly being added to the atmosphere from natural processes, such as volcanic eruptions and the respiration of animals and plants, and removed from the atmosphere by natural processes, such as photosynthesis and absorption by surface waters in lakes, rivers, and oceans. This ongoing flow of carbon through the environment is known as the *carbon cycle*, and has been responsible for the fluctuations in the levels of atmospheric carbon dioxide for hundreds of thousands of years.

Atmospheric carbon dioxide levels have historically varied from around 180 parts per million (ppm) to around 300 ppm. This means that if we were to pack a box with one million marbles (with 780,000 purple marbles representing nitrogen, 210,000 green marbles representing oxygen, and 9,820 multicolored marbles representing the other gases—such as water vapor—that are present in the atmosphere), then only 180 of them would be red marbles (representing the low end level of carbon dioxide). So a change in carbon dioxide level in the range of around 120 ppm, or 120 or so red marbles, occurred through all of the glacial and interglacial periods, when the planet's temperature was cold enough to bring about prolonged periods of cooling during an ice age, and warm enough to bring about shorter periods of warming that melted a lot of the ice and caused increased sea levels. That seemingly small change in the atmospheric carbon dioxide level coincides with a really big change in the earth's climate! Let's put it like this: If you had to count those one million molecules in the air, or the total number of marbles in the box, at the rate of one molecule, or marble, every second, it would take 16,661.7 minutes (about 11 1/2 days nonstop) until you came to the last 300 molecules (or red marbles representing the high carbon dioxide level), which you would count in the last five minutes of that 11 1/2 day period! The concentration of carbon dioxide, when compared to the total amount of molecules in the air, is very small indeed, but it can have a powerful effect on energy patterns on the earth. It is interesting to note that the carbon dioxide level in the atmosphere has increased by around 100 ppm in the last 200 years.

Methane

Methane gas is actually about 20 times better at absorbing infrared radiation than carbon dioxide, but its atmospheric concentration is a lot lower, and it does not remain in the atmosphere for as long as carbon dioxide. In fact, the atmospheric concentration of methane is often measured not in parts per million but parts per billion (ppb). The preindustrial levels of methane in the atmosphere were less than

850 ppb (or 0.85 ppm). Carbon dioxide levels at that time were around 280 ppm. So carbon dioxide has a greater contribution to the greenhouse effect than methane, due to its higher concentration in the atmosphere, even though methane is better at absorbing heat energy. Methane is a main component of natural gas. It can also enter the atmosphere from the decomposition of dead organic matter at the bottom of swamps, wetlands, and bogs, where there is little oxygen, and also from the metabolic processes of certain bacteria that are found in the digestive systems of animals. Insects such as termites also emit methane as they go about their business of eating dead wood. Methane—it's a natural (swamp) gas!

Nitrous Oxide

Nitrous oxide is produced naturally by, for example, the microbial processes in soil and water, and its preindustrial atmospheric concentration was around 280 ppb. This is lower than the concentrations of both methane and carbon dioxide at that time.

Ozone

Ozone is a naturally occurring gas that is found mainly in a layer in the stratosphere, an upper part of the atmosphere. Here the ozone can be considered to be a very good thing, because it filters out and protects us from some of the harmful effects of ultraviolet light. A very small amount of this stratospheric ozone has the potential to diffuse into the upper region of the troposphere where it can act as a greenhouse gas. Ozone in the troposphere can be considered a bad thing.

So Global Warming Is a Bad Thing, Right?

WRONG! Without the greenhouse effect all of the heat released from land and ocean surfaces would return directly back into space and the earth would get very cold indeed— down to an average temperature of zero degrees Fahrenheit (−18° C)! But because of the greenhouse effect the earth's average temperature is presently around 58°F (14.5°C). A certain amount of solar energy needs to remain in the earth's atmosphere to act as a natural heating system. Water vapor provides an easy-to-observe example. On a cold winter's night when the sky is dark and the stars are out the temperature can get very low. But if there were clouds in the sky and the stars could not be seen, the nighttime temperature would be warmer, because the water vapor in the clouds traps some of the heat. The clouds act like a blanket. During the day cloud cover has a cooling effect, because the sun's rays are blocked from reaching the earth's surface, reflected back into space from the upper surface of the cloud. On a sunny day, clouds act like a sunshade.

So Global Warming Is a Good Thing, Right?

WRONG!
Like the saying goes
"too much of a good thing is bad."
This is the double-edged sword of the greenhouse effect and the resulting global warming. We do need a certain quantity of greenhouse gases in the atmosphere in order for the earth to be hospitable for life. Up until the Industrial Revolution the levels of these gases fluctuated up and down around a mean. There was a certain periodic balance in the system that only changed over long periods of time. The problem now, according to Bill Chameides of Environmental Defense, is that "things are happening a lot faster than anyone predicted." The natural greenhouse effect is a good thing and helps to regulate the planetary temperature. Nighttime cloud cover is like a blanket around the earth, in much the same way a blanket on your bed keeps you warm on a chilly night, and you get a good night's sleep. Current levels of global warming are more like lots of extra blankets around the earth, causing it to get hotter and hotter. If you put lots of extra blankets on your bed, you would get so hot and uncomfortable that you would not sleep well and wake up and kick off some of those extra layers. You would then cool down to a temperature that enabled you to stay warm and get some sleep! Right now we have realized that we have put too many blankets

around the earth and it is getting too hot, which is not a good thing. We need to look at ways to stop putting any more blankets around the earth, and if at all possible take some off.

Some media reports have suggested that since global warming is a naturally occurring phenomenon and it keeps the planet warm, it must be a good thing and we should not be concerned about the rising levels of gases such as carbon dioxide. (Some people have even suggested that since plants need it to grow then carbon dioxide should be regarded as an atmospheric fertilizer. However, the US Supreme Court ruled that global warming emissions such as carbon dioxide should be considered pollutants and be under the regulation of the Clean Air Act.)

To get an idea of the effect of very high levels of carbon dioxide on the atmosphere of a planet, let's take a look at Venus. The atmosphere is mostly made up of carbon dioxide (95 percent; recall that in our atmosphere it is about 0.03 percent), and it is thick with highly acidic clouds that prevent us from seeing the planet's surface.

Although Venus is closer to the sun than the earth is, the cloud cover reflects about three-quarters of the sun's energy back into space. At the uppermost surface of the acidic clouds the temperatures are below freezing. The sunlight that does penetrate the cloud cover gets trapped by the carbon dioxide to produce a greenhouse effect large enough to keep the surface of Venus a nice and toasty temperature of roughly 900°F! Now that's what I would call too much of a good thing. Of course, Earth is not on the road to becoming Venus, but the global warming that is taking place is of great concern. So how did those greenhouse gases get so out of the control?

The Human Factor

Carbon Dioxide

The levels of carbon dioxide in the atmosphere fluctuate naturally throughout the year. This is due to the increased uptake of carbon dioxide by plants in the process of photosynthesis each spring and summer, which causes the atmospheric level of the gas to drop. In fall and winter the levels of carbon dioxide in the atmosphere rise, because the rate of photosynthesis by plants decreases during the winter months.

MOI?

The process of photosynthesis

Carbon dioxide + water + energy from the sun → glucose + oxygen

$$6\ CO_2 + 6\ H_2O + energy \rightarrow C_6H_{12}O_6 + 6\ O_2$$

Another natural process that produces carbon dioxide is the cellular respiration of animals and plants. When animals "burn" carbohydrates such as glucose to provide energy for movement and other metabolic reactions, they use up oxygen and produce carbon dioxide, which is exhaled as they breathe. It is the opposite reaction of photosynthesis.

The process of cellular respiration

glucose + oxygen → carbon dioxide + water + energy

$$C_6H_{12}O_6 + 6\ O_2 \rightarrow 6\ CO_2 + 6\ H_2O + energy$$

These annual fluctuations can be seen in the Keeling Curve, as can the overall increased levels of carbon dioxide caused by human activities since the 1950s.

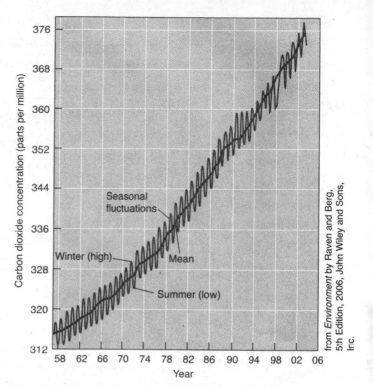

Concentration of carbon dioxide in the atmosphere, measured atop Mauna Loa, Hawaii.

from *Environment* by Raven and Berg, 5th Edition, 2006, John Wiley and Sons, Inc.

As the human population has increased, we have cleared more and more forests and woodlands to provide more places to for us to live, and also turned them into agricultural fields that provide us with food. Forests are often cleared by burning felled trees. By removing this vegetation, we remove the plants that trapped carbon dioxide through the process of photosynthesis, and the atmospheric levels of the gas increase. When the trees are burned, carbon dioxide is released through the process of combustion.

Which brings us to the major human activity that is responsible for increased global carbon dioxide levels: The burning of fossil fuels such as coal, oil, and natural gas. The carbon in fossil fuels was taken out of the prehistoric atomosphere by photosynthesizing plants, locked up and stored underground for millions of years.

Whenever we burn coal at power plants to produce electricity, drive our cars from place to place, or use the gas range to cook our food, we are taking this prehistoric carbon and releasing it into today's atmosphere as carbon dioxide. As the human population grows, we are burning more and more fossil fuels and releasing more and more carbon dioxide into the atmosphere through the combustion of these important energy sources.

The combustion of natural gas (methane)
Methane + oxygen → carbon dioxide + water + energy

$$CH_4 + 2\,O_2 \rightarrow CO_2 + 2\,H_2O + \text{energy}$$

Deforestation and burning fossil fuels are causing an imbalance in the earth's natural carbon cycle. More atmospheric carbon dioxide means more global warming.

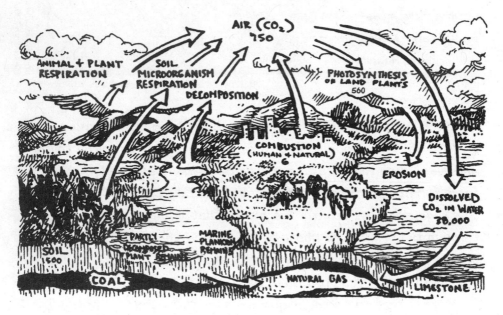

The Carbon Cycle

Most of the earth's estimated 10^{23} g of carbon is found in sedimentary rocks and fossil fuels. The values shown in the diagram represent the global carbon budget expressed as 10^{15} g of carbon. So the amount of carbon dioxide in the atmosphere is 750×10^{15} g of carbon. (That's a lot of carbon, especially when you think that one pack of Oreo cookies is around 500 g, with each cookie weighing around 10 g.)

Methane

Burning fossil fuels, such as natural gas, can release methane into the atmosphere. Methane can also escape from coal mines and leak from gas pipelines. Certain bacteria found in the digestive systems of animals release methane gas from the cellular respiration process that takes place in low-oxygen, or anaerobic, conditions. The gas is released into the atmosphere from the belching and flatulence of the animals! We raise more and more domestic animals to provide our increasing population with meat. More cows mean more burping and farting, which in turn means more methane gas getting into the atmosphere, which contributes to more global warming.

Methane is also produced naturally by the anaerobic decomposition of decaying vegetation in swamps and wetlands. Increased production of rice in paddy fields accelerates this process. Methane is also a byproduct of the decomposition of trash in landfills. The overall effect of livestock husbandry, rice farming, and landfills has caused the level of methane to reach its highest atmospheric level in hundreds of thousands of years.

Nitrous Oxide

We are increasing the levels of nitrous oxide in the atmosphere above natural conditions through increased animal husbandry (the metabolic actions of anaerobic bacteria on animal wastes produces the gas), and through the use of commercially produced inorganic fertilizers that are added to the soil in many modern agricultural operations. Nitrous oxide is also produced by many combustion processes. For example, the combustion of gasoline in cars and the burning of fossil fuels to produce electricity or to heat homes and businesses releases nitrous oxide. There has been a rapid rise in the levels of atmospheric nitrous oxide since the Industrial Revolution. When nitrous oxide passes through the troposphere a warming affect occurs, and when it reaches the stratosphere it can destroy the ozone layer.

It is interesting to look at the increase in human population since the Industrial Revolution and compare it to the increases in atmospheric carbon dioxide, methane, and nitrous oxide since that time.

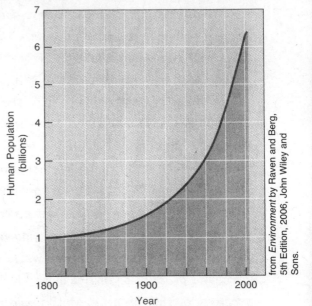

from *Environment* by Raven and Berg, 5th Edition, 2006, John Wiley and Sons.

Ground-level Ozone

Human activities are impacting the amount of ozone that is found in the troposphere at or near ground level. Once again, the combustion of fossil fuels is the culprit. Nitrous oxide and other nitrogen oxides are released by combustion, and through a series of chemical reactions a buildup of ozone in the lower part of the troposphere near ground-level results. In addition to having a role as a greenhouse gas and increasing global warming, this "bad" tropospheric ozone can have a major impact on human health and bring about other damaging environmental effects.

Water Vapor

It is very difficult to know how the levels of water vapor in the atmosphere have changed during the course of history, since water is found all over the earth. We know that sea levels rise and fall, and that there have been times when glaciers and ice sheets covered large parts of the land, all indicative of changes in the planetary water cycle. In all probability an increase in atmospheric water vapor takes place as the planet warms. This is a logical proposal: As temperatures increase, evaporation from surface water into the atmosphere also increases. Of course, if more of the heat absorbing water vapor is present in the atmosphere, then this can increase the overall greenhouse effect and impact global warming, raising the temperature which causes more water to evaporate . . . resulting in more water vapor in the atmosphere . . . increasing the greenhouse effect . . . causing more warming . . . causing more evaporation . . .

Increases in the tropospheric levels of carbon dioxide, methane, and nitrous oxide between 1850 and 2005

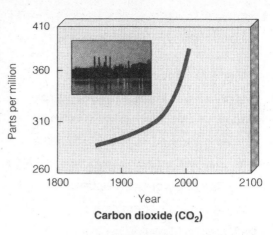

Carbon dioxide (CO_2)

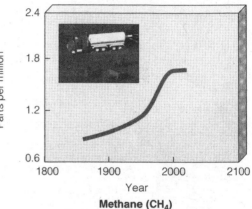

Methane (CH_4)

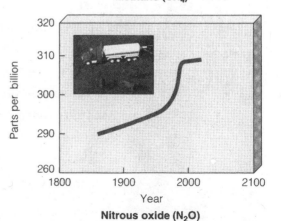

Nitrous oxide (N_2O)

from *Living in the Environment* by Miller, 15th Edition, 2007, Thomson Higher Education.

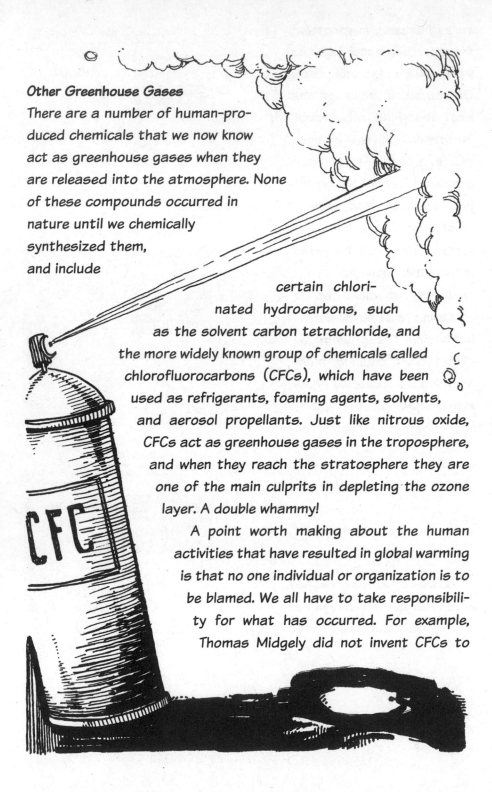

Other Greenhouse Gases

There are a number of human-produced chemicals that we now know act as greenhouse gases when they are released into the atmosphere. None of these compounds occurred in nature until we chemically synthesized them, and include certain chlorinated hydrocarbons, such as the solvent carbon tetrachloride, and the more widely known group of chemicals called chlorofluorocarbons (CFCs), which have been used as refrigerants, foaming agents, solvents, and aerosol propellants. Just like nitrous oxide, CFCs act as greenhouse gases in the troposphere, and when they reach the stratosphere they are one of the main culprits in depleting the ozone layer. A double whammy!

A point worth making about the human activities that have resulted in global warming is that no one individual or organization is to be blamed. We all have to take responsibility for what has occurred. For example, Thomas Midgely did not invent CFCs to destroy the ozone layer. He did it so that ammonia, then used as a refrigerant, could be replaced with a safer substance. CFCs fit the bill perfectly, and because people wanted to enjoy a lifestyle that included refrigeration and air conditioning, CFCs became widely used and everyone was happy. At the time, no one realized that many years later these chemicals would be causing trouble for the planet. Sometimes the effects of our actions take many years to be established and become apparent. As soon as we realized the problem with the thinning ozone layer, countries of the world, working together and through the Montreal Protocol, acted to reduce the use of, and find replacements for, these ozone-depleting chemicals. Although the concentrations of ozone in the stratosphere will take several decades to fully recover, this serves as an example of what can be done on a global scale when we all work together. We confirmed that human actions of using CFCs were having a profound negative effect on the environments of the earth and all of the species that inhabit the planet. There were few skeptics on the issue of stratospheric ozone depletion, and we witnessed a successful outcome of global environmental cooperation. As with the topic of global warming, the researchers on the ozone problem were also honored with a Nobel Prize! With

SEEMED LIKE A GOOD IDEA AT THE TIME.

regards to global warming, we are in a similar point that we once were with the ozone depletion problem. We know that human activities are causing a large enhancement in the greenhouse effect, causing temperatures of the land, air, and water to increase on a global scale. We are beginning to understand some of the consequences that could result, both now and in the future. We should not waste valuable time and energy in trying to point fingers and apportion blame. Many people travel in cars and use electricity in their homes. We all need to assess what we can do and encourage all countries of the world to once again work together and move towards addressing the global warming problem. The Kyoto Protocol is at least a good first step, albeit a difficult one to address, in that direction.

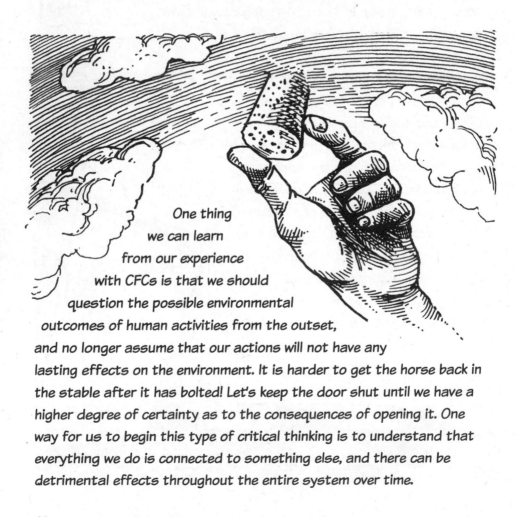

One thing we can learn from our experience with CFCs is that we should question the possible environmental outcomes of human activities from the outset, and no longer assume that our actions will not have any lasting effects on the environment. It is harder to get the horse back in the stable after it has bolted! Let's keep the door shut until we have a higher degree of certainty as to the consequences of opening it. One way for us to begin this type of critical thinking is to understand that everything we do is connected to something else, and there can be detrimental effects throughout the entire system over time.

The Earth: A Systems Approach

Throughout history many people have tried to explain the connections and balance of the natural world. From the ancient Greek idea that the universe consisted of the four elements—earth, water, air and fire—to the modern-day Gaia hypothesis proposed by British atmospheric scientist James Lovelock, who suggests that the living and nonliving components of the earth work together to maintain the conditions necessary for life through self-regulating processes. Philosophers such as Herodotus, Plato, Aristotle, and Cicero and naturalists including Jean-Baptiste Lamarck and Charles Darwin, among others, have searched for answers in order to come to a more complete understanding about how nature works, what natural balances exist, and how humans interact within nature. Some of the answers to these questions continue to be refined as we learn more about the natural world and our role in it. Some earlier ideas hold up to scrutiny and the passage of time, while others are edited or discarded as our knowledge base grows. If there are intricate connections between different components of the biosphere, then how did they occur, and how can we view them in a way that makes sense?

Increasingly, people who study the natural world have realized it is composed of a series of interacting loops and cycles that can operate independently and together. A series of systems!

Think of the human body. It is made up a number of systems that are all in operation day to day, hour to hour, minute to minute, and second to second. The systems work without conscious thought—we don't have to think about getting them to work. For example, there is the circulatory system that is responsible for pumping blood around the body, the nervous system that sends impulses from the brain to control other parts of the body, and the digestive system that is responsible for extracting nutrients and energy from food, to name just a few. To get an idea of how a system can work, let's take the regulation of body temperature as an example.

These human systems can affect one another. Heat from the metabolic processes occurring in the body can warm its temperature. Energy from the sun can also warm the body. As body heat increases, the blood warms. This is detected in the brain, and as the body gets hotter and hotter, the brain sends out a message to the skin,

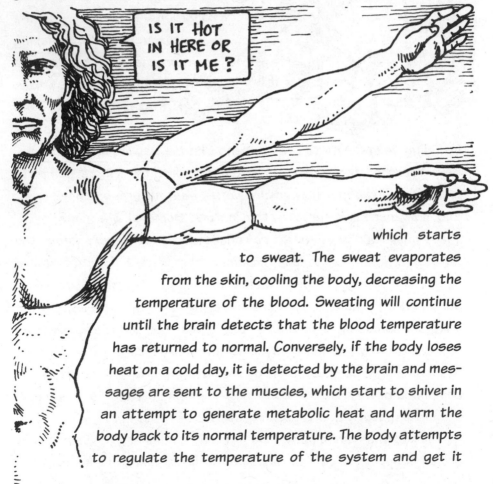

which starts to sweat. The sweat evaporates from the skin, cooling the body, decreasing the temperature of the blood. Sweating will continue until the brain detects that the blood temperature has returned to normal. Conversely, if the body loses heat on a cold day, it is detected by the brain and messages are sent to the muscles, which start to shiver in an attempt to generate metabolic heat and warm the body back to its normal temperature. The body attempts to regulate the temperature of the system and get it back to its normal, balanced state. This process is called *homeostasis*.

A similar process can be visualized by thinking of the heating and cooling system in your home. The thermostat is set to the desired temperature. If it gets too cold and the temperature drops below the set value, the heat comes on until the temperature has risen back up to the desired setting. If it gets too hot and the temperature rises above the set value, the cooling system comes on until the temperature has dropped back down to the desired setting. The temperature in the house stays near the set value, just as your body stays around the same temperature through the process of homeostasis.

So how can we define a system? In any system there are *inputs, system throughputs,* and *outputs*. In the case of the body's thermoregulation, the heat from the sun or the body's metabolism is the input; the increased heat in the body and the blood is the system throughput (also referred to as system flow), and the heat loss due to sweating is the output.

LOOPS I KNOW SUMTHIN' ABOUT.

We can also identify in a system negative and positive *feedback loops*. Negative

feedback loops tend to act to restore the balance of the system and cause it to change in the opposite direction to which it is moving. For example, as the blood temperature increases, sweating provides a negative feedback to decrease the blood temperature. It was

45

going up, the negative feedback brings it back down—body temperature has been moved in the opposite direction. Negative feedbacks tend to be a good thing in the natural world and end up having a positive outcome to the system as a whole! Positive feedback loops tend to act in a way that keeps the system changing in the same direction. For example, consider what would happen if the brain sent a message to the muscles to make us start to shiver as the temperature of the blood increased! This would cause the body to generate more heat and the blood to get warmer and warmer, causing its temperature to continually increase, or change in the same direction. We often experience such events when our body is invaded by bacteria or viruses and we develop a fever. Such examples of positive feedback show that they tend to be a bad thing in the natural world and end up having a negative outcome to the system as a whole.

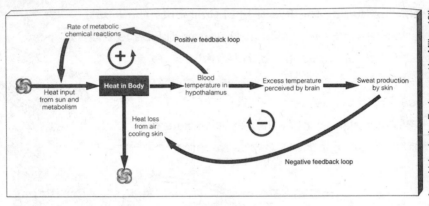

from *Living in the Environment* by Miller, 13th Edition, 2004, Thomson Higher Education.

Negative and postive feedback loops control the temperature of the human body

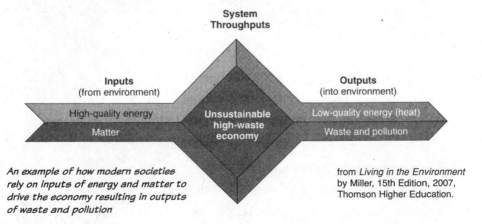

An example of how modern societies rely on inputs of energy and matter to drive the economy resulting in outputs of waste and pollution

from *Living in the Environment* by Miller, 15th Edition, 2007, Thomson Higher Education.

There are times when you can look at a positive feedback loop as being a good thing. For example, if you have a lot of money in a saving account that is gaining compound interest, the amount will keep getting bigger and bigger as the positive feedback causes change in the same direction. Your bank balance will continue to grow!

Let's use another classic example to view a system. Suppose you have a bathtub that can hold 50 gallons of water. Water could enter the bathtub by a number of inputs: you could turn on the hot tap; you could turn on the cold tap; you could pour water in using a bucket. Outputs from the bathtub could include pulling the plug or using a bucket to get rid of the water. Any method that could be used to cause water to enter the bathtub can be considered an input into the system, just as anything that causes the water to be removed from the bathtub can be considered an output. Inputs and outputs can take place quickly, like turning both taps on full or pulling the plug *and* using a bucket! Or the input and output rates could be very slow—a dripping tap or a slowly leaking plug. The water in the bathtub can be considered to be the system flow or throughput. Say we wanted to keep the bathtub at a constant level of 25 gallons. We would have to fill it to 25 gallons, and then adjust the input and output rates to be the same, so the amount of water flowing in would equal the amount of water flowing out. (Now I know that most times when you fill up a bathtub you just put in the plug and turn on the taps until the water has reached the desired depth and tempera- ture, then turn off the taps. The level

stays the same, because the input and output would both become zero. I do not suggest that anyone leave the water running straight in and out of their bathtub—it would be such a waste of water. So imagine the flowing bathtub scenario in your head!) We have set up the system so that the input from the tap equals the output from the drain—perhaps the plug is not all the way in—and the level remains at a constant 25 gallons. If there was a change of input, maybe by turning on both taps, then more water would be flowing in than flowing out and the level of water in the bathtub will rise until it overflows. A system way out of balance! In order to restore the level back to 25 gallons, we would have to increase the output to a rate at which water flowing through the drain equals the increased rate or input from the tap. This is an example of a negative feedback, because it would move the water level in the opposite direction and the system would be restored to 25 gallons and stay balanced at half-full.

If, on the other hand, we reduced the rate at which water flowed out of the bathtub, or increased the input rate by adding water to the bathtub by using a bucket, the bathtub system would get out of balance and quickly overflow.

This is an example of positive feedback,

because the water level continued to rise, or move in the same direction.

A system will remain in a balanced state until something happens to either the inputs or outputs. Consequences of such changes can result in either positive feedback, causing the system to continue to move in the direction of its already out-of-balanced state, or negative feedback, which causes the system to move back toward its previously balanced state. Such interactions are occurring in our bodies all the time, and can also be seen operating in the larger natural world.

In nature, a community of populations of different species interacting with each other and their nonliving physical environment in a certain area is referred to as an ecosystem. Together all of the earth's ecosystems interacting with each other make up the biosphere, where according to G. Tyler Miller, Jr., president of Earth Education and Research, "everything is linked to everything else." The term ecosystem can be applied to an area as small as a puddle of water or as large as mountain forest. The water cycle can be viewed as a system; the atmosphere acts as reservoir for some of the water in that system. Two inputs of water vapor into the atmosphere are evaporation from surface water (streams, lakes, rivers, oceans) and transpiration from plants. Some of the water vapor in the atmosphere results in changes in humidity at different times of the year, or may result in the formation of clouds. Outputs of water vapor from the atmosphere could be from precipitation—dew, rain, snow, and hail. The natural water cycle consists of inputs and outputs. Overall, a balance in the water level of a pond, for example, is maintained by this natural system. The level rises and falls around an already existing, normally established balanced state, depending upon the season. After a heavy rainfall the water in the pond could rise for a while, and on hot days during the summer the water level could drop, due to increased evaporation. Such fluctuations are normal in the natural scheme of things. In most cases the fluctuations are not too extreme and do not cause the pond to be in a constant flood or to dry up completely.

A series of biogeochemical cycles operate in nature that recycle chemical elements through the nonliving environment, into living things, and then back into the nonliving environment. Each one of them can be considered a system in its own right. You may have heard of some—the nitrogen cycle, the phosphorous cycle, the sulfur cycle, the water cycle, and, of course, the carbon cycle. Each of these systems serves to exemplify the close, complex interactions in the natural world between the atmosphere (air), the hydrosphere (water), and the lithosphere (earth).

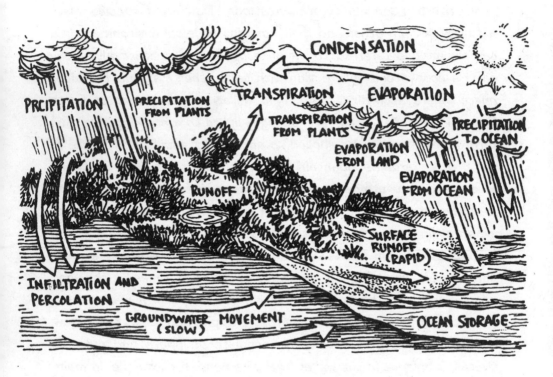

This diagram of the hydrologic or water cycle shows how water interacts with the atmosphere, lithosphere, and hydrosphere.

The Carbon Cycle: A System Out of Balance

Let's think of the carbon cycle as a system. The atmosphere acts as a reservoir of carbon, or system throughput. Inputs of carbon into the atmosphere can come from the respiration of animals, plants, and microorganisms; from naturally occurring wildfires; from burning of forests to clear land or the combustion of fossil fuels. The carbon released into the atmosphere will stay there, much of it in the form of carbon dioxide, until it is removed by a system output, such as photosynthesis, or until it transfers to another location, perhaps dissolved into the planet's surface water. Throughout the geological time period that the carbon cycle has operated it has kept levels of carbon dioxide in the atmospheric reservoir between 180 and 300 ppm. So how are we causing the carbon cycle to become a system that is increasingly out of balance?

Humans are burning more and more fossil fuels, which results in more carbon dioxide being added to the atmospheric reservoir. We have increased the input of carbon to the system. This increased carbon dioxide level has a positive feedback effect on the global climate system, increasing the earth's temperature.

Humans are also deforesting vast areas of the land all over the globe. This causes a decrease in the overall level of planetary photosynthesis and results in less carbon dioxide being removed from the atmospheric reservoir as an output. More carbon dioxide remains in the atmosphere and will have a positive feedback effect of the global climate system, increasing the earth's temperature. Of course, if the deforestation has occurred through slash and burn, as is often the case with tropical rain forests, then the carbon cycle gets hit with a double whammy: carbon dioxide input is increased by combustion, and the carbon dioxide output via photosynthesis is reduced! Such changes to the inputs and outputs of the carbon cycle system are causing the planet's biogeochemical cycles to move more and more toward an ongoing state of imbalance. Soon the level of carbon dioxide in the global atmospheric sink will surpass 400 ppm. There will be wide-reaching consequences of such a high level of this greenhouse gas, consequences that are becoming increasingly apparent in the global climate.

Chapter 3

Global Warming: The Consequences

Many of the consequences of global warming have been highlighted and discussed throughout the media, in television programs and newscasts, magazines and newspapers, radio broadcasts, blogs, Websites, and textbooks. The topic is probably the most widely publicized of any current environmental issue. Depending on the source of the information, and the particular bias of the author, some consequences have been downplayed while others have been sensationalized and exaggerated.

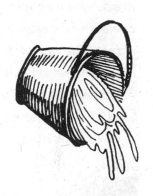

What do we know is happening now, and what could possibly occur in the future? It is clear that we are already seeing environmental changes that are a direct result of global warming. What is unclear is the extent to which these changes may continue to occur, and the impact of other changes that are predicted to take place as a result of the ongoing warming trend. Of course, the extent of any or all of the consequences will depend on how quickly the issue of greenhouse gas emissions is addressed on a global scale. More rapid action will reduce these potential outcomes, while slower action will only lead to a worsening effect. The latest report from the IPCC clearly outlines some of the potential consequences of global warming, and these conclusions have been subject to the highest level of scientific scrutiny and consensus. So what is the current state of affairs according to the latest IPCC findings?

The Temperature Is Rising

The Earth is getting warmer. We are witnessing some of the hottest years on record since the mid-1800s. The effect is global in nature with the land masses experiencing a faster increase in temperature than the oceans. The warming effect is also more pronounced farther north from the equator. The IPCC suggests that the 50-year time period from 1950 to 2000 could represent the highest temperature change in any 50-year time period in the last 500 or even 1300 years. In fact, 11 out of 12 of the warmest years in the past century and a half were recorded between 1995 and 2006. The temperature increase is behind the heat waves that have led to many deaths worldwide, such as the one that occurred in Europe during the summer of 2003. In the United States and elsewhere it seems that each year the daily temperature record for many locations is broken. The new ones that are set sometimes end up being broken again in a short time period. The years 1998 and 2005 were tied as the two hottest on record until NASA recently declared 2005 as the winner. According to the IPCC, it is likely that heat waves will become more frequent over most of the earth's land masses. During 2007 new national record highs were observed in Hungary, Slovakia, Moldova, Serbia, Macedonia, Kenya, Kuwait, Japan, Montenegro, and Bosnia. So what else is this global temperature increase bringing about?

A Global Meltdown

Snow cover on mountain ranges is decreasing. Sea ice in the Arctic and Antarctic is melting at rates even faster than some of the climate models once estimated. Ice on land masses such as Greenland and the Antarctic is melting, and glaciers are receding. Specific examples of this global meltdown include:

- ice caves in Cascades National Park, Washington

- snow cover on mountain peaks in China (Meili Snow Mountain) and Tanzania (Mount Kilimanjaro), and

- glaciers in the Himalayas (Chhota Shigri), Iceland (Breidamerkurjokull), the Swiss Alps (Tschierva), Patagonia (Upsala and Perito Moreno), Canada (Athabasca), and the United States (Boulder in Glacier National Park, Montana, and Muir and Riggs in Alaska).

Muir and Riggs
glaciers, Alaska in 1941
(Photo: William Field USGS)

Riggs Glacier, Alaska 2004
(Muir Glacier has melted)
(Photo: Bruce Molnia USGS)

"Almost all of the mountain glaciers in the world are now melting, many of them quite rapidly."
—Al Gore in the film *An Inconvenient Truth*

This melting has caused an increase in the size of glacial lakes in mountainous regions, and altered the flow in glacial and snow-fed streams and rivers. People in many parts of the world are dependent on snowmelt from mountain ranges as their predominant source of drinking water. Changes in the flow rates of glacial and snow-fed rivers will ultimately create water shortages for those populations who are dependent upon them. The Rockies supply Los Angeles and Southern California; the Himalayas supply Northern India.

Melting of sea ice has been observed in both the Northern and Southern hemispheres. In the Arctic the ice cover is shrinking, and according to the IPCC late summer sea ice may disappear completely by the end of this century.

Arctic ice cap in late summer, 1979 (Photo: NASA)

Arctic ice cap in late summer, 2003 (Photo: NASA)

Dr. Donald Perovich is one of the lead research scientists on the Seasonal Ice Zone Network project, which measures the thickness and melting speeds of Arctic ice. Based at the US government's Cold Regions Research and Engineering Laboratory (CRREL) in Hanover, New Hampshire, Perovich refers to the Arctic as "the canary in the coal mine" of global warming. He points out that in 1982 the permanent Arctic sea ice covered an area about the size of the continental United States. In 2005, the permanent ice cover had decreased by an area roughly the size of the 22 states east of the Mississippi River. "The only way you can get that much melting is by the human component," Perovich commented in July 2007. Of course the melting of the Arctic sea ice could open up the historically sought after Northwest Passage, which Perovich says would provide "a shortcut across the top of the world." The impact of this increased shipping on marine ecosystems is the subject of an ongoing study by the CRREL scientists.

The sea ice in the Arctic does not cover any land mass; it is simply a large amount of ice that floats on the upper region of the ocean. The melting of the sea ice directly affects planetary *albedo*, the amount of sunlight that is reflected back into outer space. Sunlight is reflected from the surface of the ice back into space. As the ice melts, less sunlight will be reflected and more of the sun's energy will be absorbed by the darker surface of the water. This will cause the temperature of the surface water to increase which will in turn lead to the melting of more ice. A positive feedback loop will have been established and over time less of the Arctic region will be covered by sea ice.

Ecologists are concerned about what will happen to polar bear populations should the Arctic ice continue to melt and disappear at its current rate, but younger folk have their own worries—one day the North Pole will no longer be frozen and will no longer provide a home for the jolly old elf in the red suit! Where will Santa Claus relocate when the

North Pole melts? Children all over the world look forward to his annual visit in return for being "nice" all year. Will Santa have to relocate to the South Pole as a result of global warming? I know many children who are very alarmed about the possibility of the North Pole melting. Will the reindeers drown or will they fly off in time? To these children this is a real concern. Perhaps in the future we shall witness a Million Toddler March in capital cities around the world, children carrying placards with SAVE SANTA'S HOME.

Not only is the warming in the Arctic causing the sea ice to melt, it is also having an impact on the permafrost in the tundra. Changes in these areas have been observed over the span of a human lifetime. For example, as recently as 35 or 40 years ago the tundra was frozen for more than 230 days a year; in 2005 some parts of the tundra were frozen for just 75 or 80 days. The melting permafrost in Alaska has caused land subsidence. Foundations and building walls crack, and even collapse. Telephone poles and trees lean over and eventually fall down. Shishmaref, an Alaskan island community, has been inhabited for more than four centuries. The permafrost is melting and buildings have become unsafe, causing some of the residents to contemplate evacuation from their homes or the abandonment of the whole community. As the permafrost melts, it releases trapped carbon dioxide and methane once locked up in the frozen ice.

Another example of a positive feedback loop on global warming: increased Arctic temperatures cause melting permafrost which releases carbon dioxide and methane which brings about increased Arctic temperature causing . . .

Near the other pole, an alarming event took place between late January and early March in 2002. Antarctica's Larsen B ice shelf broke up at rate that had never been witnessed before. The ice shelves in the Antarctic hold back the glaciers that cover the land mass. When the ice shelves break up, they no longer hold back the land-based glaciers. The "land ice" falls into the sea and more dire consequences can then take place.

Rising Sea Levels

As ice melts it can bring about a rise in sea levels worldwide. Water has many unique properties; one of them is that as it freezes it expands. Have you ever known anyone to suffer from burst water pipes in the winter? As the water becomes ice, it passes through a temperature where its volume increases and the pressure exerted on the pipes is strong enough to break them. When the ice becomes liquid the pipes leak as the water flows freely from them through the break. As the sea ice, icebergs, and ice shelves melt the global sea level will increase due to this thermal expansion of water. According to the IPCC, since 1993 sea levels have been rising at around 3.1 mm per year, a finding consistent with the global warming trend. The melting water from the sea ice was part of the ocean before it froze, so when it melts the impact on sea level rise will not be as great as from ice that melts from land masses that comes mainly from precipitation.

The melting of land-based ice in glaciated areas such as Greenland, Antarctica, and snow-covered mountain peaks will have a far greater impact on global sea levels than the melting of sea ice. The glacial ice in Greenland is melting at such a rapid rate that it could result in extremely high sea levels, simply by adding water to the ocean. Scientists estimate that if all the ice covering Greenland were to melt, it could raise sea levels by 21 feet worldwide!

Think of it like this—if you put an ice cube in a glass of water and measured the level of water in the glass, then let the ice cube melt and measured the level of water again, it will have gone up by a very little bit, so small that it may be hard to see. This is like what happens if sea ice or icebergs melt that are already floating in the water. Now think what would happen to the level of water in the glass if every time the ice cube melted you added another one. When that one melted you added another one and so on. After a while the level of water in the glass would Increase, and eventually water would spill over the rim. If the land ice in glaciers, or the runoff from the snow melt from mountaintops, is added to the ocean the level will keep rising as more ice, or runoff, is added. Glaciers breaking off land masses such as Greenland and the Antarctic and falling into the ocean have a major impact on sea level rise. Think of what happens when a garden pond freezes in winter and then melts in spring. The level of water does not change much, because the ice on top of the pond is part of the water that was already in the pond. If during the winter when the pond was frozen you dumped a truckload of ice cubes onto the frozen surface of the pond, when the thaw came the water level would be higher as a result of the extra water that was added to the pond ecosystem.

It is the melting of this land-based ice that is really worrying climatologists. But why? If we had more water in the seas and oceans, there would be a bigger area for marine species, right? Yes, but rising sea level has other consequences that may off-set any ecological benefits. Roughly half of the earth's human population lives in coastal areas; the potential effects of sea-level rise could be devastating and widespread. Low-lying coastal plains would become flooded and submerged by the advancing seas. Homes that were on the coast will become submerged. Homes that were miles inland will become, for a while, homes by the sea until the sea advances past them and they become submerged as well. This scenario is not too far-fetched. Perhaps the residents of Washington, DC do not realize that they live in what could become oceanfront property!

The global rise in sea level would affect many areas, including:

- The coastal United States, especially the states of Florida, Texas, North Carolina, and Louisiana

- The Netherlands

- Low-lying islands in the South Pacific and Indian Oceans (the Maldives)

- The Baltic shore of Poland, and

- The delta regions of Bangladesh.

Changing Ocean Currents

Water has unique heat-absorbing properties, and as the oceans get warmer their currents could be affected. Vertical mixing occurs naturally between layers of warmer water that lie on top of the cooler layers at the bottom of the oceans. Ocean currents are connected, forming a conveyer belt system. Warm, lower salinity water on the surface circulates from the Pacific Ocean, through the Indian Ocean, and the North and South Atlantic before cooling, becoming more salty, and sinking to return back to the Indian and Pacific Oceans in cold deep currents. The influx of freshwater from the melting sea ice and glaciers will alter the salinity of the seawater and could affect this global circulation.

We do not yet know how surface and deep ocean currents will be affected.

Some oceanographers and climatologists think that the Gulf Stream, perhaps the most well-known of these currents, could slow down. If these surface currents change, we may see a resultant change in global climate.

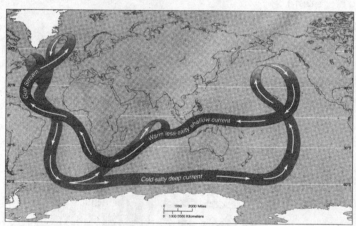

from *Environment* by Raven and Berg, 5th Edition, 2006, John Wiley and Sons, Inc.

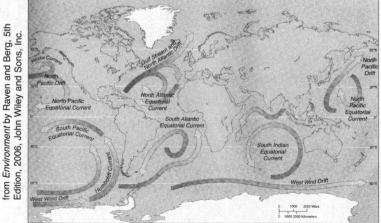

from *Environment* by Raven and Berg, 5th Edition, 2006, John Wiley and Sons, Inc.

We may also witness a change in the El Niño event, which occurs periodically and lasts between nine to twelve months. During an El Niño year, warm ocean surface water moves from the western to the eastern Pacific Ocean. As a result, there is less rainfall in parts of Asia, Indonesia, and Australia, bringing drought to these regions, and an increase in rainfall and storms along the Pacific coast of North and South America, bringing flooding and mudslides. It is anticipated that as ocean temperatures change due to global warming, the frequency and the extent of El Niño events will increase. This in turn could result in climate changes on a global scale. Some regions will become drier while others will become wetter; some regions will become cooler while others become warmer.

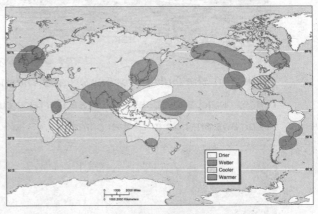

Climate patterns related to an El Niño event

from *Environment* by Raven and Berg, 5th Edition, 2006, John Wiley and Sons, Inc.

Changes in Precipitation and the Water Cycle

The water cycle is one of Earth's global systems. Powered by energy from the sun, water cycles through rainfall, surface water flow, sea currents, plant growth, and other processes all over the globe. The IPCC has sounded the alarm for what is already being observed and what could occur in the future as global warming impacts this most fundamental of the earth's systems.

By observing measurements taken between 1900 and 2005, the IPCC has ascertained that rainfall has increased in northern Europe, the eastern parts of North and South America, and parts of Central Asia. This increased rainfall has caused flooding. It has also been seen that rainfall has declined in the Mediterranean, the Sahel and southern parts of Africa, and parts of Asia. It appears the incidence of drought in these regions has increased as well.

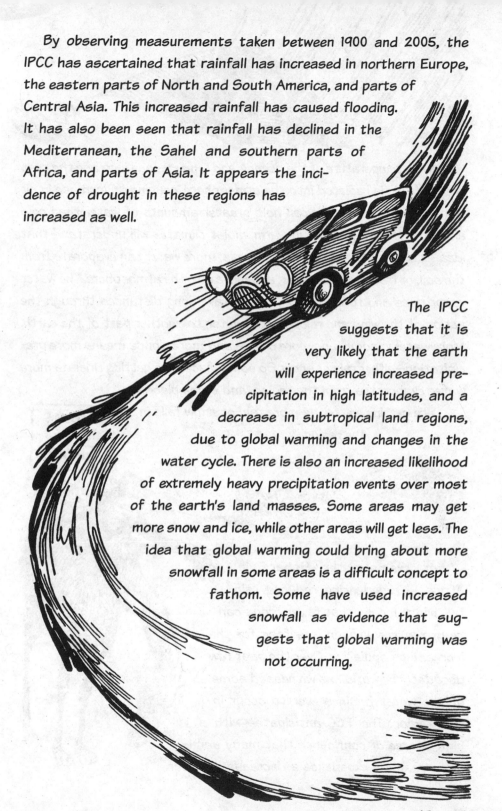

The IPCC suggests that it is very likely that the earth will experience increased precipitation in high latitudes, and a decrease in subtropical land regions, due to global warming and changes in the water cycle. There is also an increased likelihood of extremely heavy precipitation events over most of the earth's land masses. Some areas may get more snow and ice, while other areas will get less. The idea that global warming could bring about more snowfall in some areas is a difficult concept to fathom. Some have used increased snowfall as evidence that suggests that global warming was not occurring.

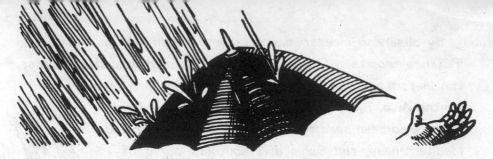

As the temperature of the oceans and land masses increase the heat they contain is radiated into the atmosphere, causing an increase in air temperature. Warmer air can hold greater amounts of water than cold air. Those of you that live in warm, moist climates will understand that idea—it's called *humidity!* In warm areas, more water can evaporate from the soil, or from surface water, and move into the atmosphere. The water vapor rises and forms clouds, which travel long distances through the action of the wind and relocate the water to another part of the earth. More water vapor means more clouds and more clouds means more pre-cipitation—either rain or snow. So some places are getting drier as more water evaporates from the region, and some places are getting wetter as more rain or snow falls in the region.

WELCOME TO THE JUNGLE.

For example, the Sahara in Africa is experiencing reduced rain-fall, which means that fewer crops can be grown, resulting in less food for the increasing population. Over the past few decades, the world has witnessed some of the worst famines ever to occur in this region. The IPCC anticipates with a high degree of confidence that many semi-arid areas will experience a decrease in water resources due to climate change brought about by global warming. Desert regions will spread, due to less rainfall and more evaporation. Changes in precipitation patterns can only lead to increased conflict over water resources in areas of the world, such as the Middle East, that already have limited water supplies. Decreased rainfall and higher incidence of drought means more wars over water.

Many areas have experienced an increase in the frequency and intensity of flooding. Summer 2005 brought devastating floods to parts of Europe. Parts of Asia are witnessing increased flooding, and during July 2007, parts of England experienced some of the worst flooding in that country's history.

This flooding and drought that is brought about by changes in the water cycle can also lead to some confusion. For example, how can an area like California be in one of the worst droughts in its history, then a few months later experience some of its worst floods? Why doesn't the rainwater ease the drought? Why does it instead cause mudslides and other problems? Many factors come into play. Because we have altered the land surface through activities such as housing developments and road construction, when it rains, a lot of the water runs off a hard surface, rather than infiltrating the soil and replenishing the groundwater. The water table drops as there is less input of groundwater; wells dry up or have to be dug deeper, and in some areas water conservation measures have to be enacted like banning the watering of lawns and washing cars. Then when it rains the water runs off into the gutters, roads, streams, and rivers, resulting in flooding. This is exacerbated if a lot of rain falls in a short period of time. Too much rain cannot be handled satisfactorily by a municipality's storm water drainage system. Sometimes water saturates the ground, causing it to become unstable and slide. The floodwater eventually flows back to the ocean through streams and rivers as it travels on its way through the water cycle. So even during a period of drought we can have flooding, which does little to provide long-term relief from the drought!

In order for humans to have a constant water supply, we have had to look to other parts of the water cycle, tapping into the resources of rivers and lakes. Rivers have been diverted to provide water for human consumption and irrigation purposes. The effects of our actions can be seen in even the greatest rivers —the Indus,

the Nile, the Colorado, and the Niger; and in bodies of water such as Asia's Aral and Caspian seas, lakes Chad and Malawi in Africa, and North America's Mono and Great Salt lakes. Although the changes that can be seen in these rivers, lakes, and seas can be mostly attributed to overdrawing and water diversion projects, we can only speculate the extent to which global warming will compound the problems in the future. As the water temperatures in rivers and lakes increase, water quality and the thermal structure of these aquatic ecosystems may be affected. Many temperate lakes and ponds are subject to a semiannual turnover each fall and spring—a vertical mixing of warm and cold layers—which is an important process in the life of a lake or pond. The effect that global warming may have is yet to be fully recognized.

The water cycle is in part responsible for the weather that we experience on a daily basis and the climate that we experience in our region over the longer term. Monsoons and tropical storms may become more intense. There has been an increase in tropical storm activity in the North Atlantic Ocean. We are also observing changes in wind intensity and wind patterns. As the oceans get warmer and radiate greater amounts of heat to the atmosphere, the strength and energy of ocean storms may increase. We are familiar with the names of the hurricanes that formed in the Atlantic Ocean recently—Andrew, Mitch, Ivan, Emily, and of course 2005's devastating Katrina. We may not be as familiar with the Pacific Ocean's typhoons—Amang in 2007 and Chanchu in 2006—or the cyclones of the Indian Ocean—Akash and Gonu in 2007. Currently, the IPCC cannot establish any clear link between global warming and increased storm frequency and intensity; it is a difficult systems analysis problem. Dr. Kerry Emanuel of the Massachusetts Institute of Technology has evaluated the data, and he believes that "hurricanes are lasting longer and getting stronger because of global warming." Other climatologists argue that there is insufficient data to conclusively make that claim. What is clear, however, is that increased global temperatures affect the water cycle causing changes in precipitation patterns worldwide, which could in turn affect the severity of some weather events.

The Effect on Ecosystems and Biodiversity

Just as our body's metabolism takes place within an optimal temperature range, so do many of the natural world's cycles and systems. When the temperature rises too high, or drops too low, there could be far-reaching consequences within an ecosystem. We are already observing some effects of global warming on species and ecosystems. Through our knowledge of how ecosystems operate, we can use logical, critical thinking to make predictions as to other consequences that may take place as global warming continues.

There are many things that a plant or animal species requires for its ongoing survival: heat, water, food, minerals, oxygen, carbon dioxide, acidic or basic conditions. If all of the parameters are at satisfactory levels then the organism thrives. However, if only one of the parameters falls above or below the optimal range, the species will be adversely affected. The parameter that is not at the optimal level, thereby limiting a species' population, is known as the *limiting factor*. For example, a shortage of water during a period of drought can cause a decrease in the number of both plants and animals in an ecosystem. This then has an effect on the biodiversity of that ecosystem.

Biodiversity refers to the numbers and types of all of the species in an area. If a limiting factor causes a decrease in the numbers of ladybugs, a predator for example, there could be an increase in the number of aphids (which eat plants and are eaten in turn by ladybugs), and increased damage to plants within that ecosystem. When an area begins to lose some of its biodiversity, the whole ecosystem suffers. The interdependence of all species within the ecosystem is controlled by a number of limiting factors which could act as a system input or output.

The limiting factor could be a living organism (parasite, predator, or

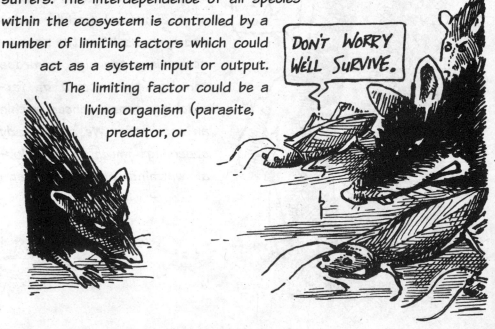

DON'T WORRY WE'LL SURVIVE.

disease causing bacteria) or a nonliving factor (water, temperature, acidity, pollution, or pesticide). If a limiting factor takes one thing out, like the ladybug, other species, like the aphid, will be affected, which in turn can have a knock-on effect on plants in other parts of the ecosystem. For example, the Asian ladybug has been introduced into North America (input) and it is competing with native species of ladybugs for food and habitat. This is causing a decline in the native ladybug populations (output). If you took one of the spark plugs from your car's engine, the car would not operate at its optimal level. Eventually other things may occur to the engine as a result and cause the car to break down completely. Increased global temperature can have similar detrimental consequences in many terrestrial and aquatic ecosystems. The IPCC has determined with a high degree of confidence that

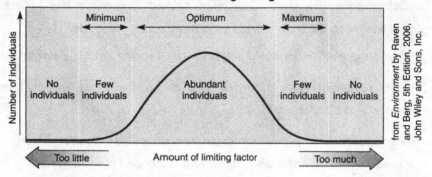

Too much or too little of any environmental resource can affect the number of species in a population

"20–30 percent of assessed species are likely to be at increased risk of extinction . . . and could be as high as 40–70 percent if global temperatures continue to rise." In Costa Rica the disappearance of the golden toad has been linked to global warming.

In some parts of the world people have observed an earlier timing of spring events. The movie *An Inconvenient Truth* examined the shifting seasons in the Netherlands, which has had an effect on some bird and caterpillar populations that are dependent on one another. In Great Britain some butterfly species have disappeared from some areas and are now found in areas that they previously did not inhabit. In the United States increased damage to trees by the pine bark

REST IN PEACE.

beetle has been noted. In Yellowstone National Park, a decrease in trees due to increased beetle infestation as a result of the warmer climate may have an effect on the population of grizzly bears, which depends on the trees as part of its food source. Increased temperatures have been linked to the decrease of mangrove trees in parts of the Florida Everglades.

Other concerns that have been raised regarding the effect of global warming on ecosystems and biodiversity include:

• Swamp, marsh, and wetland habitats may become drier in some areas.

• Drier soils may result in increased incidence of wildfires in some forest and woodland areas.

• Agricultural crop production could be reduced as a result of heat stress on plants, water shortages, and increased plant diseases.

• Some poisonous plants, such as poison ivy, may become more toxic.

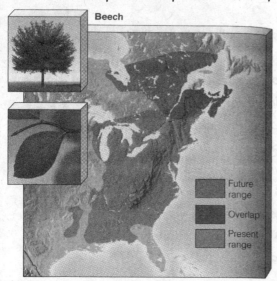

Beech

Future range

Overlap

Present range

• The ranges of some plant and animal species may change. For example, researchers at the University of Minnesota have suggested that beech trees, which are common throughout eastern North America, may in the future only be found in the northern part of Maine and the southeastern region of Canada.

from *Living in the Environment* by Miller, 15th Edition, 2007, Thomson Higher Education.

The full extent of any or all of these consequences on terrestrial ecosystems remains to be seen; much depends on how quickly and effectively the human race responds to the challenges of global warming and greenhouse gas emissions.

There is already concern about the effect decreasing Arctic ice coverage will have on the population of polar bears that use the ice packs for travel and hunting. Obviously if the ice sheets decline or disappear altogether then this must have a detrimental effect on any species that use them as an important part to their existence. Maybe we will observe a change in the polar bears' range? A similar concern has been voiced regarding the penguin population in the Antarctic. Some penguin species depend on a covering of sea ice as an important aspect of their breeding cycle. If the Antarctic sea ice continues to melt and break off from the mainland, we can only speculate the effect that this will have on the penguin numbers. It is unlikely to be beneficial to them.

Rising water temperatures in marine and freshwater ecosystems have also resulted in changes in their natural balance and biodiversity. In addition to changes in ice cover, there have also been alterations in salinity, and of dissolved oxygen and carbon dioxide levels. Concern has been raised regarding the future productivity of the world's fisheries, which are already threatened by overfishing. Declining populations of a number of algae and plankton species have been noted. An algal bloom can result in dead zones in aquatic ecosystems, as was observed in the Baltic Sea in 2005. As algae dies and drifts to the bottom of a body of water, it provides food for decomposers. As the decomposers feed on the dead algae they consume oxygen and release carbon dioxide through the process of cell respiration. The water's oxygen level can fall so low that it is not sufficient to support aerobic life—think fish—and the area essentially becomes a dead zone. In addition to changes in water temperature, algal blooms can be triggered by increased inputs of nitrate and phosphate nutrients contained in runoff from agricultural regions. Phosphorus is often a limiting factor in algae growth; if there is phosphorus in abundance, there is abundant algae too. The aquatic ecosystems of the Chesapeake Bay, the largest estuary in North America, have been detrimentally affected by dead zones caused by algal blooms.

Algae also grow on the underside of sea ice and are an important

food source for some marine species. A decrease in the amount of sea ice also means less algae and therefore less food for those species that depend upon them. It is likely that we will observe a domino effect throughout the whole marine food chain in some areas. There are already alarmingly decreasing krill populations in the

Antarctic Ocean. This may be the result of human activities. If the Antarctic sea ice continues to melt as the water temperature increases then this will have a dramatic affect on the base of the Antarctic food chain. Less algae will be produced, which means less food for the krill, which means less krill, which means less food for the other marine species that depend on the krill, such as the blue whale. Again, it is evident that a disruption to one part of the system can have widespread consequences.

Many of the world's most pristine coral reefs have been under threat for a long time as a result of pollution and other human activities. Increasing ocean temperatures may only compound the problem. Coral reefs are often referred to as the "rainforests of the ocean" and their high biological productivity is of great importance to many species of marine life. Coral consists of an animal, the coral polyp, and a plant, zooxanthallae, living in close harmony with each other in a mutually beneficial relationship. The coral polyp provides the zooxanthellae with a protected place to live, and the zooxanthallae provides the polyp with food produced by photosynthesis. (That's why coral reefs are only found in the upper surface of the ocean in the region where light can penetrate, called the *euphotic zone.*) Coral provides an example of an important, finely-tuned interaction between two different species within the marine ecosystem. Both species need each other for survival. *Coral bleaching* occurs when the beautifully colored coral becomes a ghostly white—the coral has died. If the water temperature gets too warm and becomes out of the range of tolerance for the zooxanthallae they are expelled from the coral. Global warming may be just another human-induced detriment to affect the coral reefs and the other marine species that are dependent upon them for their continued existence.

Effects on Ocean Chemistry

An idea to emerge over the past few years is the concern that the acidity of the seas and oceans could increase due to the higher levels of dissolved carbon dioxide that occur in warmer waters. This hypothesis is being investigated through ongoing scientific studies; it is a rational expectation based on current knowledge. We know that when carbon dioxide dissolves in water it forms carbonic acid, so we could draw a conclusion that more dissolved carbon dioxide leads to more carbonic acid which could then react with and affect the shells of marine species. The increased acidity could affect the calcium metabolism of shell-forming marine life. It may also impact the skeletal formation of bony fishes. It is interesting to note that in the 1960s DDT was found to affect the calcium metabolism and therefore eggshell formation in birds like the bald eagle and the osprey. This finding resulted in banning the use of DDT in 1972 and protection of the bald eagle through the Endangered Species Act.

The Effect on Humans

Humans rely on the natural capital provided by the earth for many activities at all times of the year. In recent generations we seem to have become disconnected with the natural world and have withdrawn from nature, not fully appreciating its wonder as we once did. This argument is eloquently expressed in Richard Louv's *Last Child in the*

Woods. Even so, many people still go out into the natural world and participate in many activities—hunting, skiing, snowboarding, bird-watching, snowmobiling, snow shoeing, horseback riding, photography, hiking, leaf-peeping, camping, ice fishing, fishing . . . the list goes on! What could global warming do to limit the recreational activities that we love to participate in?

Not only could global warming have an effect on the aesthetic pleasure we receive from nature, but in many instances it could have economic consequences as well. For example, if a region that relies on skiing as part of its winter tourism activity receives less winter snow and more winter rainfall, then people will not come to the area to ski. The whole economy—from ski resorts to hotels and restaurants—will not have a good year financially. Businesses could close after a few years of low snowfall, particularly if the trend becomes a permanent feature and consequence of global warming's effect on climate. Ski seasons could become shorter, and in fact all mountain sports could be affected.

The northeastern United States and Canada have a thriving maple syrup industry that is dependent upon cold winters for the sap to run well in the spring thaw. Warm winters are already impacting the amount of maple syrup that has been produced in recent years.

Another important component of the region's economy occurs well before the sap starts to run. In the fall many people visit these areas to marvel at the spectacular fall kaleidoscope as the maples and other trees change color. Changes in regional climate may affect the fall foliage, which in turn will impact tourism and the local economies. And if the maple trees' range changes in a pattern similar to that predicted for the beech tree (see page 73) in many areas there won't even be a maple to see!

Global warming has begun to have an impact on human health. It is likely that we will see changes in infectious diseases in some places. As the temperature increases, the range of disease-carrying or disease-causing insect vectors could change. For example, the mosquito and tsetse fly are now found living at higher altitudes in some parts of Africa. A warming globe could bring about the spread of diseases such as malaria that prior to the temperature changes were not a problem in certain regions. This is of particular concern if you happen to be a resident of Harare, Zimbabwe, or Nairobi, Kenya, as these two cities are at elevations that mosquitoes previously did not inhabit. Now these cities are at risk due to the migration of mosquitoes to the higher altitudes as the temperature warms. Time will tell how widespread the changes in infectious diseases will become. Higher carbon dioxide levels have been implicated in increased amounts of ragweed pollen. This could lead to detrimental effects for those who are susceptible to allergies and prone to respiratory distress caused by asthma.

The entire human population is vulnerable to the threats posed by climate change brought about by global warming, because everyone is susceptible to the effects of drought, flood, heat wave, disease, and famine. No one is immune from the risks posed by global warming.

Will Global Cooling Help Offset the Consequences of Global Warming?

We know that substances in the atmosphere, such as dust, smoke, soot, and aerosols can reflect sunlight back into space and as a result cool the planet. This phenomenon was apparent in the aftermath of the eruption of Mount Pinatubo in 1991. James Hansen, a scientist at NASA, put together a climate model that suggested the earth would initially cool due to the smoke and ash discharged into the atmosphere from the volcanic explosion. After a period of time he predicted that the earth would then warm up and return to its pre-eruption temperature. His climate model was found to be correct as global temperatures were monitored over the succeeding years.

Some particles can also absorb sunlight and radiate heat toward the earth's surface. There are also natural processes that occur in the atmosphere that clean the air of pollutants through chemical reactions. Some of the positive and negative feedback mechanisms that operate in the atmosphere are complex and cause scientists to argue as to the extent that any global cooling effect may have.

Professor Richard Lindzen, an atmospheric scientist at the Massachusetts Institute of Technology, believes that aerosols in the atmosphere will actually make the earth cooler in 20 years than it is today, and that global warming as a result of human-induced effects is exaggerated.

Professor Alan Thorpe, a renowned meteorologist and head of the Natural Environmental Research Council believes otherwise. "Although there are significant uncertainties in certain aspects of the [system] models, particularly in the effects of clouds, there was no reason to suppose that the models have a systematic bias towards human-induced global warming. If anything the models have underestimated the degree of warming," he suggests.

More research is warranted in the area of global cooling and its potential impact on global warming. In fact, the global warming that traps heat in the troposphere causes the stratosphere to get colder, which results in an increased breakdown of the ozone layer by CFCs! Such is the interconnection of the atmospheric systems.

One thing is certain; the overall trend is that the earth is warming despite any impact that global cooling is having on the system as a whole. It is probably not a good idea to pollute the atmosphere with more dust, smoke, or soot in the hope that these particles will help cool down the earth! Nor would it be logical to increase sulfate particles (that lead to acid rain) just because they may have a cooling effect in the atmosphere. There are other, more sustainable ways to reduce the extent of global warming in the future.

The IPCC has summarized the examples of impacts associated with global average temperature change in the table below.

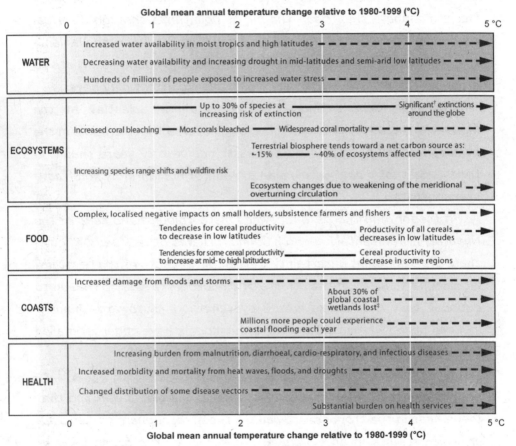

Global mean annual temperature change relative to 1980-1999 (°C)

† Significant is defined here as more than 40%.
‡ Based on average rate of sea level rise of 4.2 mm/year from 2000 to 2080.

Chapter 4

Global Warming: The Solutions

One of the pitfalls of environmental education is that it can leave people with a sense of hopelessness. All we tend to hear about are the doom-and-gloom scenarios that will occur if the human race continues with business as usual. Rather than take the ostrich approach and bury our heads in the sand in the hope that the problem will go away on its own, I urge people to first understand what the problem is, then to determine what action we all can take to alleviate the problem. Admittedly, some problems are more difficult to address than others, but nevertheless address them we should. Some approaches may fail, but out of this can come learning, development of new problem-solving ideas, and some leaps forward in finding solutions.

There are many examples throughout history of earth-shattering advances in science, engineering, medicine, and technology that have come about through many years of arduous work, failures, and small gains. We have unraveled the mysteries of atomic structure, designed machines that can fly at supersonic speeds at high altitudes and that explore the depths of the ocean, cured or alleviated diseases that once were lethal, probed the outer limits of the universe, and sent men to the moon. We have come together to take global action to protect the stratospheric ozone layer. Now is the time for the entire human race to come together and take action on the problem of climate change and global warming. This is the defining issue of our age, and we all need to rise to the task of finding solutions. At no other time in history has it been so important for every-one to understand the environmental impacts of human activities on the earth's systems. Systems that are out of balance are unsustainable and ultimately collapse. Why is it important to learn about environmental systems and sustainability? Because in most regions of the world, lives will depend on it. We all need to step up to the plate and restore balance to the global systems that control the earth's temperature and climate. If humans use the problem-solving creativity that we have shown in the past, these solutions are within our grasp. The plea for searching for these global solutions was expressed succinctly by Rajendra Pachauri, chairman of the IPCC, in his acceptance speech for the Nobel Peace Prize in Oslo on December 10, 2007.

Neglect in protecting our heritage of natural resources could prove extremely harmful for the human race and for all species that share common space on planet Earth. Indeed, there are many lessons in human history which provide adequate warning about the chaos and destruction that could take place if we remain guilty of myopic indifference to the progressive

DON'T JUDGE— I AM LOOKING FOR A SOLUTION.

erosion and decline of nature's resources. . . . In recent years several groups have studied the link between climate and security. These have raised the threat of dramatic population migration, conflict, and war over water and other resources as well as a realignment of power among nations. Some also highlight the possibility of rising tensions between rich and poor nations, health problems caused particularly by water shortages, and crop failures as well as concerns over nuclear proliferation.

Climate change and other consequences of global warming are a current and future reality; they should not simply be viewed by us as a gloomy situation, but as a unique opportunity for global cooperation on an unprecedented scale. Pachauri shares a similar optimistic

perspective:

The implications of these changes, if they were to occur, would be grave and disastrous. However, it is within the reach of human society to meet these threats. The impacts of climate change can be limited by suitable adaptation measures and stringent mitigation of greenhouse gas emissions . . . the global community needs to coordinate a far more proactive effort towards implementing adaptation measures in the most vulnerable communities and systems in the world. There is substantial potential for the mitigation of global greenhouse gas emissions over the coming decades that could offset the projected growth of global emissions or reduce emissions below current levels.

He ended his Nobel Prize acceptance speech with the following question: "Will those responsible for decisions in the field of climate change at the global level listen to the voice of science and knowledge, which is now loud and clear?"

The Role of the Governments of the World

Undoubtedly the development of policies by each nation's government will be a difficult process. It will take the global cooperation of all governments to work together to address global warming, which is causing widespread, long-lasting global effects. The political cooperation will need to be long-term and be subject to compromise between the developed and developing nations. Some government policies will involve incentives and others will provide subsidies. We will need governments to support both the use of current technology and the development of new technologies. Some solutions may involve controversial actions, such as moving away from our overconsumption of fossil fuels and developing new renewable sources of energy. A lot of the argument against such measures lies in the fear of potential damaging economic consequences. However, Lester R. Brown, president of the Earth Policy Institute, says, "The challenge is to build a new economy—at wartime speed. The good news is that we have the technologies to do it."

There has been a call for governments to implement a tax on carbon in order to move toward a lower carbon economy. Others suggest that an energy tax should be added to the cost of fossil fuels. Many people feel that governments themselves should set an example by incorporating energy and fuel efficiency strategies into all aspects of their operations, as oftentimes all branches of government are responsible for a very large amount of a country's total energy consumption. Other sectors of society feel that governments should provide tax incentives for business and industries that engage in energy efficiency protocols, and reduce or remove tax benefits for those businesses and industries that use fossil fuel for their operations.

The effects of global warming will not be felt equally by all peoples. For example, parts of Africa and Asia, as well as small island nations, will be the first to experience some of the detrimental impacts and consequences of global warming. Many of these nations are in the developing world. There is an ethical and moral argument for the developed world to help these poorer countries to move toward economic and environmentally sustainable practices through sharing technological and energy efficiency strategies. It is morally imperative that ethics, fairness, and equality be at the core of the global policies that are adopted. It is the view of Rajendra Pachauri that "every country in the world has to be committed to a shared vision and set of common goals and actions that will help us all move towards lower levels of emissions of carbon dioxide." In the words of environmentalist G. Tyler Miller, Jr., president of Earth Education and Research, "This is one of the world's greatest ethical challenges. In 2006, after a meeting with climate scientists, 86 leaders in the US Christian evangelical movement came out with a joint statement, calling for federal legislation requiring cost-effective, market-based reductions in carbon dioxide emissions. They also quoted the Bible on the need to help our poorest global neighbors who will be hardest hit by climate change." It is apparent that people in all walks of life are coming to realize the moral underpinnings of the need for concerted action on a global scale in the race to address the problem of global warming. This crosses all ethnic and religious boundaries.

The call is out to the governments of the world to work together and tackle the issues that global warming raises. All citizens of the world need to raise their voices and express their concerns to those officials that represent them in their respective governments at the local, regional, and national levels. We all need to speak with one voice. We need to move away from the conflict, wars, famine, and instability that environmental degradation can bring to a country's people. The turbulence in Darfur is a tragic example of such devastation. In an editorial article in the *Washington Post* in June 2005, UN Secretary-General Ban Ki-moon stated that the slaughter in Darfur was triggered by global climate change. "The Darfur conflict began as an ecological crisis, arising at least in part from climate change. UN statistics showed that rainfall declined some 40 percent over the past two decades, as a rise in Indian Ocean temperatures disrupted monsoons. This suggests that the drying of sub-Saharan Africa derives, to some degree, from man-made global warming. It is no accident that the violence in Darfur erupted during the drought. When Darfur's land was rich, black farmers welcomed Arab herders and shared their water. With the drought, however, farmers fenced in their land to prevent overgrazing. For the first time in memory, there was no longer enough food and water for all. Fighting broke out. Any real solution to Darfur's troubles involves sustained economic development, perhaps using new technologies, genetically modified grains or irrigation, while bettering health, education, and sanitation." According to Eric Reeves, a Sudan researcher at Smith College, the number of dead from the conflict in Darfur may be close to 500,000.

By addressing such issues that affect us all globally, we can bring about stabilization and peace to many parts of the world currently in turmoil. On a global scale, Pachauri suggests that

> peace can be defined as security and the secure access
> to resources that are essential for living. A disruption in
> such access could prove disruptive of peace. In this
> regard, climate change will have several implications, as
> numerous adverse impacts are expected for some popula-
> tions in terms of access to clean water, access to suffi-
> cient food, stable health conditions, ecosystem resources,
> and security of settlements.

Is it not the right of everyone to expect nothing less from their government? Many people think so. In his recent book *Plan B 3.0: Mobilizing to Save Civilization,* Lester R. Brown outlines such a strategy. "*Plan B 3.0* is a comprehensive plan for reversing the trends that are undermining civilization. Its four overriding goals are climate stabilization, population stabilization, poverty eradication, and restoration of the earth's ecosystems." The centerpiece of the strategy "is an ambitious plan to cut carbon dioxide emissions 80 percent by 2020 in order to hold the world temperature rise to a minimum. This will be achieved by a full-court press to raise energy efficiency worldwide, expand the earth's forest cover, and develop renewable sources of energy in order to phase out *all* coal-fired power plants." No doubt Plan B 3.0 will have its critics and its supporters. Former President Bill Clinton comments, "Lester Brown tells us how to build a more just world and save the planet . . . in a practical, straightforward way. We should all heed his advice."

Some governments are indeed stepping up to the plate and announcing actions that will help reduce carbon dioxide emissions. The European Union has pledged to cut greenhouse gas emissions to 20 percent below 1990 levels by the year 2020. The United Kingdom announced that all new schools will become zero-carbon emitters by 2016. On December 3, 2007, the prime minister of Australia, Kevin Rudd, in the first official act of his newly elected government, signed a document ratifying the Kyoto Protocol committing Australia to the agreement. Furthermore, Rudd set a target for his country to reduce greenhouse gas emissions by 60 percent from 2000 levels by 2050. The country's Department of Climate Change released a statement saying, "Australia will participate actively and constructively in the negotiations working towards a post-2012 agreement which is equitable and effective. Our position is that any binding commitments need to encompass both developed and developing countries if we are to be successful in tackling climate change." In a ceremony in Indonesia on December 11, 2007, Prime Minister Rudd handed over the documents to UN Secretary-General Ban Ki-moon, saying, "This has been a decision of our government, a decision taken on the first day of our government in office, so it is a great honor to present this to you." He added, "Climate change is the defining challenge of our generation," and mentioned that the next global agreement must yield pledges from both developed and developing countries. "We expect all developed countries, *all developed countries*, to embrace a further set of binding emissions targets, and we need developing countries to play their part, with specific commitments to action."

It remains to be seen what actions other governments around the world will take and how quickly the actions will be implemented. The next pivotal meeting is scheduled to take place in Copenhagen in 2009. As of January 1, 2008, the United States stands alone among industrialized nations—the only country that has yet to ratify the Kyoto Protocol and make a pact with other world governments to take a stand against global warming. The US also remains the world's top greenhouse gas emitter.

On February 16, 2005, the Kyoto Protocol treaty became law in the 141 countries around the world that had signed it. On the same day, Greg Nickels, mayor of Seattle, Washington, launched the US Mayors Climate Protection Agreement. This grassroots initiative asks cities from around the country to commit to three things:

- Meet or beat the Kyoto Protocol targets in their own communities through a variety of approaches

- Urge the state and federal governments to enact policies to meet or beat the suggested Kyoto Protocol target for the United States (a seven percent reduction from 1990 levels by 2012)

- Urge the US Congress to pass bipartisan greenhouse gas reduction legislation, which would establish a national trading system.

By June 2005, 141 mayors had signed on, and in May 2007, Tulsa mayor Kathy Taylor became the 500th mayor to add her city to the list. As of December 26, 2007, a total of 754 mayors from cities in all 50 US states, plus Washington, DC, and Puerto Rico, had signed the Mayors Climate Protection Agreement. These city governments represent nearly 77 million US citizens, roughly one quarter of the US population. This broad support clearly indicates that the US is moving forward and taking action in regard to greenhouse gas emissions at the local level. At the federal level action has stagnated.

So how do countries and world regions compare? What percentage does each contribute to total greenhouse gas production? According to *An Inconvenient Truth*, the answer is:

United States	30.3%
Europe	27.7%
Russia	13.7%
Southeast Asia, India, and China	12.2%
Central and South America	3.8%
Japan	3.7%
Middle East	2.6%
Africa	2.5%
Canada and Greenland	2.3%
Australia	1.1%

The atmosphere, in addition to being a very complex system, can also be regarded as a global commons that everyone on Earth shares. The pollution that one country emits will in time affect plants, animals, and humans around the globe.

"Imagine all the people sharing all the world." —John Lennon

Solutions for Dealing with Global Warming

As Rajendra Pachauri points out, the way to the deal with the global warming problem is with strategies that involve both mitigation and adaptation. *Mitigation strategies* incorporate measures to reduce the level of greenhouse gas emissions; *adaptation strategies* are those that devise ways to reduce the detrimental effects of global warming already taking place and prepare for future impacts. Both strategies will play important roles in the years ahead, and the time to begin implementation is *now.* So what can be done to slow the rate of global warming?

Mitigation strategies include taking steps to improve energy efficiency, reducing fossil fuel consumption, moving toward using carbon-free renewable energy, and storing or sequestering carbon dioxide in the soil, vegetation, underground, and in the oceans.

Improving Energy Efficiency

We will need to look at all aspects of energy consumption in every sector of society and implement measures that will increase the overall efficiency with which the energy is used. For many years, Amory Lovins at the Rocky Mountain Institute has been a proponent of reducing energy consumption through increased energy efficiency. The headquarters of his institute in Colorado serves as a model for what can be done by using energy-efficient design and sustainable living practices. The heating bill for the building is less than five dollars a month! Lovins believes that such measures can be taken by the United States and other countries to reduce energy waste.

So from what sources does the world get its energy? Most of it comes from nonrenewable fossil fuels.

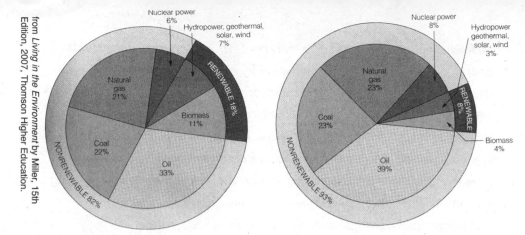

from *Living in the Environment* by Miller, 15th Edition, 2007, Thomson Higher Education.

In the United States, the energy is used by the approximate amounts in the following sectors.

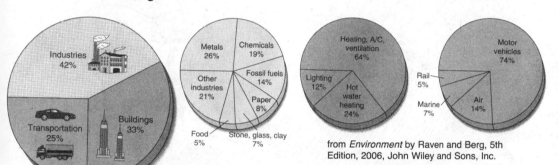

from *Environment* by Raven and Berg, 5th Edition, 2006, John Wiley and Sons, Inc.

More than 12 percent of the energy used in American buildings is for lighting. But do we need to use so much 24/7/365? What about the rest of the world?

"Earth at Night," November 27, 2000. Keeping the light on for us!
Credit: C. Mayhew & R. Simmon (NASA/GSFC), NOAA/ NGDC, DMSP Digital Archive

Reducing energy waste is referred to as *energy conservation*. Energy conservation can be achieved by making devices that use less energy but do the same amount of work. The US Environmental Protection Agency and the US Department of Energy have partnered in a government supported initiative, Energy Star, that helps industry, businesses, and individuals to use energy efficient products that save money, save energy, and protect the environment by reducing the amount of greenhouse gas emissions.

Energy Star ratings can be found on a wide range of items that are used in the home and at work.

Appliances, heating and cooling systems, electronics, lighting, food preparation and cooking supplies, and office equipment products can be purchased that come with the government's Energy Star seal of approval. Ceiling fans are now 50 percent more efficient than older models, as are refrigerators. Most electrical devices are now more energy efficient than ever before. This saves money for the consumer and is good for the environment. Did you know that:

- Replacing all exit signs in US buildings with Energy Star–approved signs would save $75 million a year in energy costs.

- If one in ten homes used Energy Star appliances, it would be like planting 1.7 million acres of trees.

• Changing one incandescent light bulb in every US home with an Energy Star light bulb would save enough energy to light more than 3 million homes a year and save $600 million in energy costs and prevent the equivalent greenhouse gas emissions from over 800,000 cars.

• Replacing the five most-used incandescent light bulbs in every US household with Energy Star light bulbs would save $8 billion in electricity costs each year and be the equivalent of reducing the greenhouse gas emissions from 10 million cars.

• Replacing all room air-conditioning units with Energy Star–approved units would prevent the emission of 1.2 million pounds of greenhouse gas emissions, the equivalent of greenhouse gas emissions from 100,000 cars.

Buildings can be better insulated as an energy efficiency measure. In fact, anything that can be done to make our homes, businesses, and industries more energy efficient results in less consumption of fossil fuels such coal, oil, and natural gas. Reduced consumption can be achieved directly, by using less heating oil in the winter, or indirectly, by using less electricity, which means burning less coal in power plants. Around half of the electricity in the US is generated from coal-fired power plants, saving electricity can add up to significant reductions in carbon dioxide emissions. Many large corporations and businesses are moving to more sustainable operating methods through introducing energy efficiency measures.

There is a lot of potential for the construction of more energy efficient "green" buildings in the future. The new headquarters of the

Chesapeake Bay Foundation in Maryland and the Environmental Studies Center at Oberlin College in Ohio are two excellent examples of such construction. The US Green Building Council has established the Leadership in Energy and Environmental Design (LEED) rating system, which is regarded as the benchmark for the design, construction, and operation of high-efficiency green buildings. More and more new construction projects are being designed in order to receive LEED certification. Some even include "cooling" roof gardens!

Some people feel that being "green" is a concept that we need to rethink and make stronger, to incorporate such initiatives into all aspects of our lives in order to achieve sustainability and protect the environment in the future. In April 2007, Thomas Friedman, a Pulitzer Prize–winning columnist for the *New York Times*, had this to say in an article entitled "The Power of Green":

> In the world of ideas, to name something is to own it. If
> you can name an issue, you can own the issue. One thing
> that always struck me about the term "green" was the
> degree to which, for so many years, it was defined by its
> opponents—by the people who wanted to disparage it.
> And they defined it as "liberal," "tree-hugging," "sissy,"
> "girlie-man," "unpatriotic," "vaguely French." Well, I want to
> rename "green." I want to rename it geostrategic, geoeco-
> nomic, capitalistic and patriotic. I want to do that
> because I think that living, working, designing, manufactur-
> ing and projecting America in a green way can be the basis
> of a new unifying political movement for the 21st century.
> A redefined, broader and more muscular green ideology is
> not meant to trump the traditional Republican and
> Democratic agendas but rather to bridge them when it
> comes to addressing the three major issues facing every
> American today: jobs, temperature and terrorism. Because
> a new green ideology, properly defined, has the power to
> mobilize liberals and conservatives, evangelicals and athe-

ists, big business and environmentalists around an agenda that can both pull us together and propel us forward. That's why I say: We don't just need the first black president. We need the first green president. We don't just need the first woman president. We need the first environmental president. We don't just need a president who has been toughened by years as a prisoner of war but a president who is tough enough to level with the American people about the profound economic, geopolitical and climate threats posed by our addiction to oil—and to offer a real plan to reduce our dependence on fossil fuels. The next president will have to rally us with a green patriotism. Hence my motto: Green is the new red, white and blue.

So apart from improvements in energy efficiency, what other ways are there to become more patriotic—more green?

Reducing Fossil Fuel Consumption

The use of coal to generate electricity wastes about two thirds of the energy in the coal itself. Even electricity generated by nuclear power loses more than 80 percent of the energy in the nuclear fuel pellets. Looking at ways to make these processes more efficient in the future will be beneficial. Establishing more cogeneration power plants will enable a greater utilization of energy from the respective fuel source. For example, a power plant that uses oil as a means to boil water to generate the steam that turns the turbine to produce the electricity could then send the steam through a heat exchanger that transfers the energy from the steam to heat nearby buildings. This is an example of cogeneration: both electricity and heat are produced and utilized from the power plant.

In addition to the energy used in operating buildings and industries, another major use of fossil fuel is in transportation, a large amount being used specifically in motor vehicles. The US Congress recently passed legislation to increase the fuel efficiency of vehicles to 35 miles per gallon by 2020. Let's put this into perspective. In 1908, when the Ford Model T was introduced to the American people, it got a respectable 25 mpg. Admittedly, cars have got bigger and faster since then, but to think that the current fuel efficiency average in America is less than that of the Ford Model T, and may only be ten miles per gallon more than the Model T by 2020, should give one pause. Does this mean that we have not had the insight and ingenuity in America to design and manufacture more fuel-efficient cars than we drove more than 100 years ago? I do not think so! We do have vehicles today that are capable of more than 50 mpg, but the overall fleet average is still below 25! We need to apply the same high levels of creativity and problem-solving that were called upon during the space race in the 1960s. When President John F. Kennedy announced that the United States would have a man on the moon by the end of the decade, it triggered a flurry of innovation in science, engineering, and technology not usually seen in peacetime. This was and still is a remarkable achievement that culminated in Neil Armstrong's famous words as he stepped onto the surface of the moon on July 20, 1969: "This is one small step for man, one giant leap for mankind."

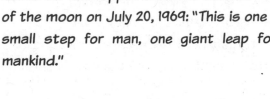

What do you think might have happened to the automobile fuel efficiency if a previous president had called for 75 mpg by the end of a particular decade during the 20th century? What if a future president calls for 100 mpg by 2020? Could we do it? I believe so.

Another defining moment in the space program came when the crippled Apollo 13 spacecraft was safely piloted to Earth. Many problems had to be solved to accomplish that task. One in particular was to devise an apparatus that would reduce the amount of carbon dioxide in the air the astronauts were breathing. The trick: building the device with only those items already on board the spacecraft. Scientists stared the problem in the face and solved it. Failure was not an option for those on the ground or in the spacecraft. What if we applied the same sense of urgency, ingenuity, creativity, and problem-solving to reducing the amount of carbon dioxide emitted into the air by motor vehicles? I believe that we have people in the country with the intelligence and fortitude to do just that!

YEAH, I MAYBE COULD REDUCE A LITTLE.

We are now seeing a change in the way that new vehicles are being advertised. The amount of fuel they use is mentioned in most of the television commercials, although the values bragged about are mainly in the 30 to 35 mpg range. With the ever-increasing price of gasoline, the American consumer should be looking for fuel efficiency in a motor vehicle. After all, the fleet averages in Japan and China are already more than 45 and 35 mpg respectively, and the European Union has set a target of at least 52 mpg by

2012. In testimony to the US Senate Committee on Energy and Natural Resources on January 30, 2007, David L. Greene of the Oak Ridge National Laboratory stated, "the consumption of oil produces additional costs that are of great significance to us as a nation but are generally not considered by individuals in their car purchase decisions: economic costs of oil dependence; military, strategic, and foreign policy costs of oil dependence; climate change impacts of carbon dioxide emissions and other environmental impacts."

It is time to look deeper into those new-car advertisements and make the link between better gas mileages and reduced global warming. It is time for a president to throw down the gauntlet and challenge the auto industry to vastly increase the fuel economy of motor vehicles in the short term. It is time for consumers, all of us, to demand it.

The importance of the link between climate and the economy was evident during the first week of January 2008. Bad weather caused the closure of oil exporting ports in Mexico. The effect was soon felt on Wall Street. The cost of a barrel of oil teetered around new record highs of 100 dollars. The higher the cost of a barrel of oil, the higher the gas prices at the pumps, and the more the consumer, you and I, pay to fill up our cars. So bad weather in this case means less money in our wallets. Now this is *not* to say that the bad weather that closed the ports was due to global warming. But it does make one wonder about extreme weather events predicted to occur as a result of global warming, and on the negative impacts to our economy.

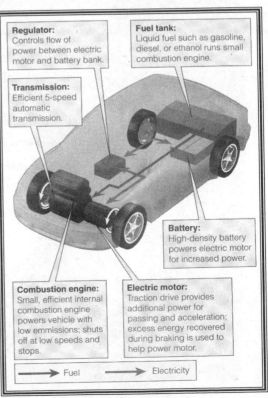

Features of a fuel efficient hybrid car

Regulator: Controls flow of power between electric motor and battery bank.

Fuel tank: Liquid fuel such as gasoline, diesel, or ethanol runs small combustion engine.

Transmission: Efficient 5-speed automatic transmission.

Battery: High-density battery powers electric motor for increased power.

Combustion engine: Small, efficient internal combustion engine powers vehicle with low emmissions; shuts off at low speeds and stops.

Electric motor: Traction drive provides additional power for passing and acceleration; excess energy recovered during braking is used to help power motor.

→ Fuel → Electricity

from *Living in the Environment* by Miller, 15th Edition, 2007, Thomson Higher Education.

It seems to be common sense to drive fuel-efficient motor vehicles.

Vehicles are being developed that not only use gasoline more economically, but use alternative fuels. A growing number of hybrid vehicles get better mileage than the gas-only counterparts.

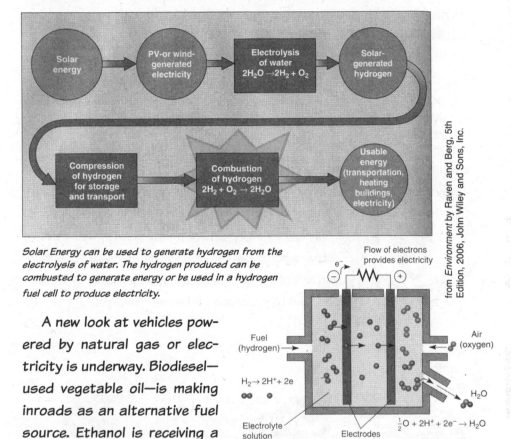

from *Environment* by Raven and Berg, 5th Edition, 2006, John Wiley and Sons, Inc.

Solar Energy can be used to generate hydrogen from the electrolysis of water. The hydrogen produced can be combusted to generate energy or be used in a hydrogen fuel cell to produce electricity.

A new look at vehicles powered by natural gas or electricity is underway. Biodiesel—used vegetable oil—is making inroads as an alternative fuel source. Ethanol is receiving a lot of attention as a cleaner burning fuel. The jury is still out as to the full potential of each of these options. Ethanol in particular may not be a panacea. Farmers are getting government subsidies to grow crops that can be used in ethanol production. This not only will increase the prices of some other foods, but could lead to negative environmental impacts from increased water use for irrigation and greater use of fertilizer, which could result in nutrient enrichment of aquatic water systems.

Currently, vehicles powered by hydrogen fuel cells are being developed and tested in some areas. Again, time will tell as to the impact that they may have. If the hydrogen that is used in the fuel cells is produced from oil, as most of it is now, then we will still be ultimately dependent on fossil fuel, even though hydrogen is cleaner to burn in a combustion engine or a fuel cell. When hydrogen is burned with oxygen it produces water vapor as a byproduct, *not* the major greenhouse gas carbon dioxide.

If we could turn our creative scientific minds to producing large quantities of hydrogen from the photoelectrolysis of water—if we could use sunlight to chemically split water molecules into hydrogen and oxygen—now that would be a giant leap for mankind. The challenge is to find the catalyst that would bring about this reaction. Nature has already solved this challenge: In the plant world photosynthesis is catalyzed by the plant pigment chlorophyll. Part of the sequence of reactions involves a similar water splitting process. A world based on hydrogen fuel is predicted by some futurists.

Body attachments
Mechanical locks that secure the body to the chassis

Air system management

Universal docking connection
Connects the chassis with the drive-by-wire system in the body

Fuel-cell stack
Converts hydrogen fuel into electricity

Rear crush zone
Absorbs crash energy

Drive-by-wire system controls

Cabin heating unit

Side-mounted radiators
Release heat generated by the fuel cell, vehicle electronics, and wheel motors

Hydrogen fuel tanks

Front crush zone
Absorbs crash energy

Electric wheel motors
Provide four-wheel drive; have built-in brakes

Features of a prototype car designed by General Motors and powered by hydrogen fuel cells

from *Environment* by Raven and Berg, 5th Edition, 2006, John Wiley and Sons, Inc.

Moving toward Renewable Energy Sources

Before the US Congress passed the new energy bill in December 2007, a provision that would have required utility companies to produce 15 percent of their generated energy from renewable energy sources was removed from the legislation. What is a renewable energy source? What potential do they really hold for the future?

Any resource on Earth that can be replaced in a short time period by natural processes, as long as it is not being used faster than it is being replenished, can be viewed as renewable. Examples of renewable energy sources include sunlight, water, wind, biomass, and geothermal heat. If hydrogen could be produced from the electrolysis of water via solar power or wind-generated electricity, it too would fall into the renewable energy category. A potential benefit of switching away from fossil fuels to using renewable sources would be a significant reduction in carbon dioxide emissions. Switching to renewable fuels would therefore have the positive environmental impact of slowing the rate of global warming.

Solar energy is free—that's the upside. The downside is that not everywhere receives the same amount of sunlight on a daily, seasonal, or yearly basis. However, it is a promising alternative.

Solar energy can be used to heat homes either passively or actively.

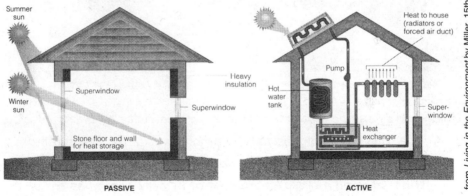

from *Living in the Environment* by Miller, 15th Edition, 2007, Thomson Higher Education.

Passive solar heating takes into account the placement and construction of a building and relies directly on the sun's energy for heat. In the future architects may design many new buildings to take advantage of passive solar design, particularly those that aim to secure LEED certification.

Active solar heating can be incorporated into a new or existing building. Sunlight is transferred into the structure using a heat exchanger.

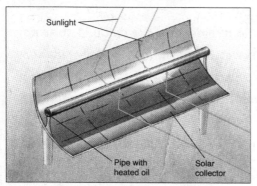

OH, YEAH, I'M GOOD.

Active and passive solar design can be implemented together, depending on location, and both have advantages and disadvantages in terms of cost, maintenance and repair, and ease of installation.

Another way that solar energy is being used is in solar thermal plants. Reflectors heat oil, and that heat energy is transferred for other uses. For example, the superheated oil can be used to produce steam to drive a turbine that generates electricity. Some solar thermal plants use large arrays of mirrors to reflect sunlight and focus it on a central receiver, known as a *power tower*.

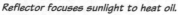
Reflector focuses sunlight to heat oil.

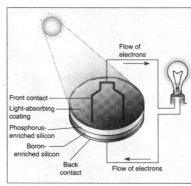
A typical PV cell

from *Environment* by Raven and Berg, 5th Edition, 2006, John Wiley and Sons, Inc.

Solar energy can be converted directly to electricity by *photovoltaic (PV) cells*. This technology is particularly beneficial for use in remote areas of developing countries and has widespread application in the industrialized world as well. As the cost of PV cells continues to come down their use will increase. For example, the State Assembly in California passed legislation that would reduce greenhouse gas emissions by 25 percent by 2020. This has spurred a growth in the solar energy business, creating several thousand jobs in research and development, production, and installation of solar panels on roofs. It is not surprising that three quarters of the US demand for solar energy comes from homeowners and companies in California.

Nancy Floyd, managing director of Nth Power, says that "through innovation and volume, prices are coming down." The hope for California is that one day electricity from solar power will be less costly than from coal or natural gas power plants.

Electricity can also be generated by the flow of *water* through a hydroelectric dam. This can be done on a large scale—the Hoover Dam in the American Southwest and, of more recent vintage, the Three Gorges Dam in China are both examples—or on a much smaller scale by using micro hydro turbines that can be easily located in rivers or streams. Hydropower has benefits and drawbacks.

I'M BACK.

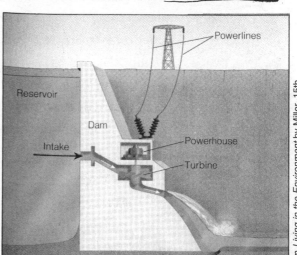

from *Living in the Environment* by Miller, 15th Edition, 2007, Thomson Higher Education.

from *Living in the Environment* by Miller, 15th Edition, 2007, Thomson Higher Education.

Harnessing the energy from *wind* is a growing industry worldwide and has a large potential for expansion. Wind turbines can be located anywhere that wind blows at a fairly constant rate. Many turbines can be grouped together into a *wind farm*. They are cost efficient, can be installed quickly, and do not emit carbon dioxide. They can be located in fields that can still be used to grow crops or graze livestock. A number of projects are being located offshore to take advantage of the ocean winds. There is opposition to wind energy; some people do not like the look of the turbines, earlier models used to be noisy, and the turbine blades may kill birds that fly into them. It should be pointed out that most bird deaths in the United States are the result of pet cat attacks; destruction of natural habitat also plays a role.

WIND FARM— HA!

We can also use *biomass*—plant materials or animal waste—to produce energy. Solid materials that can be burned include wood logs and pellets, wastes from cropland and timber harvesting, animal waste, and combustible municipal wastes. There are waste-to-energy incinerators in many locations; municipalities separate and burn their waste, rather than dump it into a landfill, and electricity is produced. Wood-chip-fired power plants also provide electricity in some locations. Manure can be used to generate biogas, which contains methane. In addition, liquid fuels can be derived from biomass, namely ethanol, biodiesel, and methanol.

Heat energy that is found in the earth's warm interior—*geothermal energy*—can be harnessed. Buildings and homes can be heated and cooled using a geothermal heat pump. Electricity can be generated as

Ooooooo, I MAKE MYSELF Sooo HOT!

well. Although some carbon dioxide emissions are associated with the use of geothermal energy, it is much less than the carbon dioxide emissions from fossil fuel. One drawback to the use of geothermal energy is the lack of suitable locations for its widespread use. However, Iceland is looking to use geothermally-generated electricity to electrolyze water in order to produce hydrogen to fuel its fishing fleet in the future.

So there are a number of alternative renewable energy sources. Each country can assess its own situation and decide which source might be potentially useful as a fossil fuel replacement. In the future, electricity may be generated from a decentralized power system, with smaller power plants using a variety of energy sources to generate electricity and feed it into the grid system. There is much for governments, businesses, and the industry leaders to contemplate as they work together to slow global warming by using more renewable energy sources.

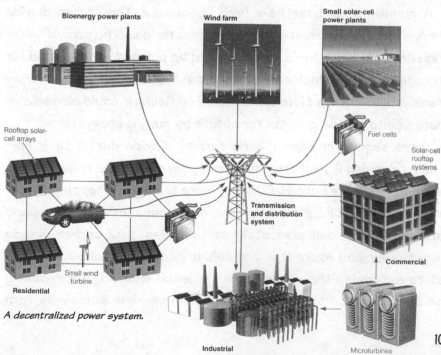

A decentralized power system.

from *Living in the Environment* by Miller, 15th Edition, 2007, Thomson Higher Education.

107

Storing or Sequestering Carbon Dioxide

One way to reduce the level of carbon dioxide emissions into the atmosphere is to capture the gas before it is emitted. Capturing carbon dioxide from a smokestack, for example, would make burning coal a more environmentally friendly energy source. The technology exists, but it is expensive to implement. But the rewards could be great, as coal is the most prevalent of all fossil fuels worldwide. There are chemicals, such as calcium and potassium hydroxide, that can absorb carbon dioxide and remove it from an effluent gas stream. Similar liquid scrubbers can be, and are currently, used to remove sulfur dioxide from smokestack emissions in an attempt to reduce the amount of acid-rain-causing substances in the atmosphere. But once the carbon dioxide has been removed from the exhaust flow, what can be done with it? Where can it be stored or sequestered for the long term?

A number of options have been suggested. The carbon dioxide removed from a smokestack or other emitter could be stored under pressure or in some liquid form. It could be pumped underground for storage in porous rock seams, such as sandstone, chalk, or limestone, or even pumped into old depleted oil fields. It could also be discharged into the deep ocean for uptake by marine ecosystems.

Trees sequester atmospheric carbon dioxide during their lifetimes, releasing it back into the atmosphere when the trees die and decompose or are burned. Planting more trees (or stopping deforestation) would therefore help control the amount of atmospheric carbon dioxide. Some plants, like switchgrass, take carbon dioxide from the air and store it in the soil. It can be cultivated. We have yet to determine the effectiveness of such methods for reducing the amount of carbon dioxide in the atmosphere; some may turn out to be cost prohibitive or not feasible from a long-term ecological or sustainability perspective.

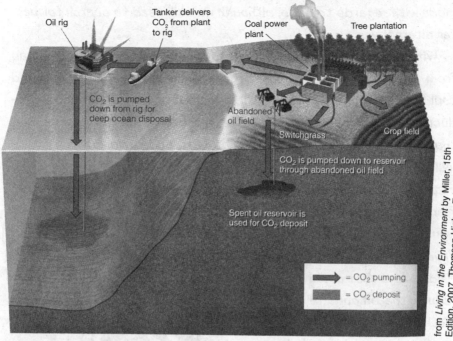

from *Living in the Environment* by Miller, 15th Edition, 2007, Thomson Higher Education.

Mitigation methods for removing and storing carbon dioxide

As part of the solution to global warming, and in addition to mitigation strategies, adaptation strategies are needed that devise ways to reduce the detrimental effects already taking place and prepare for future impacts, as Rajendra Pachauri points out in AR 4.

According to G. Tyler Miller, Jr., president of Earth Education and Research, some examples of the adaptation strategies for responding to the long-term effects of climate change are shown in the figure below:

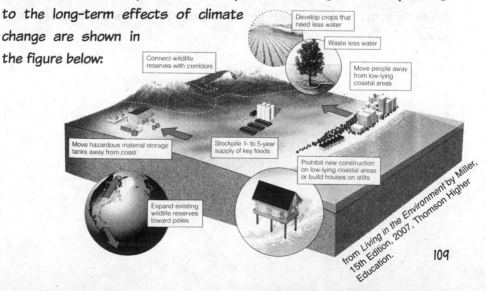

from *Living in the Environment* by Miller, 15th Edition, 2007, Thomson Higher Education.

Reducing the Levels of Greenhouse Gases Other than Carbon Dioxide

Since the Montreal Protocol was written in 1987, inroads have been made with regards to CFCs, although we still need to search for better alternatives.

With regard to methane, increased measures to capture the gas from landfills should be implemented on a wider scale. A method could be developed to retrieve the gas that is vented during oil production, rather than burning it off, a process called *flaring*. Ensuring that natural gas pipelines are free of leaks and installing leak detection systems would reduce the amount of methane released into the atmosphere. Planting high-yield varieties of rice would reduce the methane produced from paddy fields. Ways that could reduce the amount of methane produced by livestock are also being assessed and investigated.

The agricultural community could take a leading role in reducing the amounts of nitrous oxide that enters the atmosphere. Using slow-release fertilizers and reducing the amounts of nitrogenous fertilizers that are put onto fields are strategies that farmers and large agribusinesses can adopt.

Chapter 5

Global Warming: What Steps Can I Take?

No one can do everything, but everyone can do something; the question to ask yourself is, What can I do to help reduce my emissions of greenhouse gases, in particular carbon dioxide? Now some people, no matter what they hear about the human-induced global warming, will choose not to believe in the science, whether published in an IPCC report or the local newspaper. Some will tell you that they remember 20 or 30 years ago, when scientists were suggesting that the earth was cooling and could enter another ice age. "That never happened, so why should I believe the scientists who are now telling me the earth is warming? It will probably not happen either." Well, global warming *is* happening. For those people, believing the science, giving up their gas-guzzling vehicles that get a whopping ten miles per gallon, giving up other aspects of their energy-consuming lifestyle, is too much. Just like they will probably not put any credence into the fact that record winter high temperatures were recorded at a number of locations in Virginia on January 7, 2008 (Richmond, Blacksburg, Danville, Roanoke, and Dulles Airport). (Although these records could just be new one-day highs, they may suggest that the temperatures in 2008 might exceed those in 2005, the hottest year on record.) Fortunately, the folks in this category are becoming the minority.

So why has it been such an uphill battle for a lot of people to understand and accept the science behind the global warming phenomenon? Granted there has been a lot of bias and misinformation, but the reasons go deeper than that. When you think of a scientist, what do you picture in your mind? Chances are it is a white male in a white lab coat with a head full of funky wild white hair, clipboard in hand, and you cannot understand a word that he says! If you want to see what I believe a scientist looks like, look in the mirror. Anyone can be a scientist! Unfortunately, in many instances we are not educating our young people to actually *think* like scientists. Despite a push for more inquiry-based approaches, a lot of education in the sciences relies on the memorization of facts in an attempt to cover all of the material that is required for any hope of success on standardized tests. That is one of the reasons that American students lag behind students in many other countries when it comes to assessments in high school science ability. Not everyone learns best through rote memorization, and there are multitudes of learning styles in every classroom in the country that can better be served by differentiated instruction. If we want to train students to become scientifically literate adults, then we have to treat them like scientists when they are at school, and not just ask them to regurgitate textbook facts, or have them perform simple cookbook lab experiments as a means of scientific study. We need to devise ways to introduce differentiated instruction that stimulates critical thinking as a method of scientific inquiry. Students should be challenged to design their own lab experiments to answer a real-life question, rather than giving them off-the-shelf labs that offer limited opportunities for critical analysis.

To me, it is like the difference between a cook and a chef. Anyone can follow a recipe and bake a cake. Sometimes it is as simple as opening a box and adding a certain amount of milk or number of eggs! The cake doesn't always look like the one on the box, but at least the set of instructions have been followed and the experiment performed to bake the cake. Others might bake the cake from scratch with a recipe from a cookbook. We can all be cooks in our own way. Chefs, by contrast, are those individuals who do not simply follow a recipe, but modify it, play around, and experiment, even create a new cake of their own.

The same goes for scientific experimentation. If we ask students to think of the experiments themselves, to design them, to perform them, and, just like scientists, modify them to become better able to explain and support their hypotheses, then we are creating scientific thinking strategies in the minds of today's students. Sadly, a lot of science education in the United States during the past few decades has not created enough chefs. In an effort to appease standardized tests we have even moved away from creating cooks; we simply want a student to memorize a recipe, sometimes without even being given the opportunity to bake the cake. It is no surprise to me that people have a hard time accepting and understanding even the most basic scientific analysis supporting global warming. They simply have not been given the opportunity to critically examine data from an early age. This could be rectified through a different approach to education: an inquiry-based approach.

Do Try This at Home—If You Can!

I challenged my class of high school environmental science students to design an experiment that they could use to convince me that carbon dioxide does indeed affect the temperature of the air, and at the same time demonstrate the greenhouse effect.

What would you suggest in response to my challenge?

The students worked individually, and then in groups, explaining their ideas to each other and brainstorming. I helped insomuch as I facilitated this creative process. The students then conducted a series of experiments that they had designed. Each time they analyzed the data and the experimental procedures, learning from their failures as well as successes, until they came up with what I would say was their definitive experiment.

What exactly did they do?

They took three empty, clear-plastic, two-liter soda bottles and removed the labels. Into one bottle they added 100 ml of water, into another 100 ml of potassium hydroxide, and into the third 100ml of water and one quarter of an Alka-Seltzer tablet. (The fizz that occurs when Alka-Seltzer is put into water is bubbles of carbon dioxide.) They inserted a thermometer into a rubber stopper that fit into the top of each bottle. The stoppers were inserted into each of the three soda bottles, which were then placed on a table in the middle of an array

of four 250-kilowatt heat lamps; a fourth thermometer was placed on the table before the lamps were switched on. The temperature in each of the bottles and the reading from the thermometer on the table was recorded every 30 seconds for a period of five minutes.

So what were the results of this simple experiment? The bottle with water represented the carbon dioxide level that is normally present in the air. The bottle that had the Alka-Seltzer contained more carbon dioxide than is present in the atmosphere. The bottle that contained the potassium hydroxide solution would have less carbon dioxide than the normal atmosphere, because some of the gas would be absorbed by the solution. (Calcium hydroxide solution could be used; it absorbs carbon dioxide as well.) The temperature increases were calculated for each bottle and the students graphed the results.

The effect of carbon dioxide on air temperature

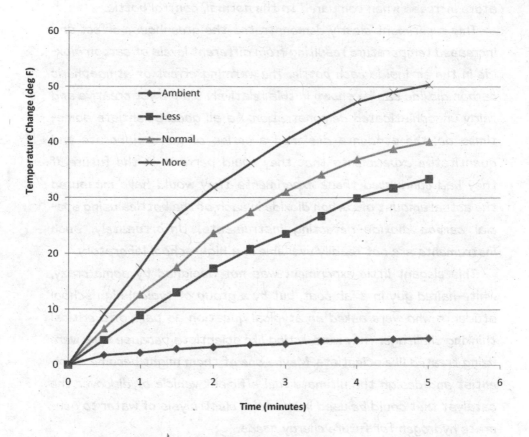

What do these results show? The thermometer on the table that was not enclosed in a bottle showed that the air temperature around the bottles increased by nearly 5°F during the five-minute period (showed by the curve labeled "ambient"). The bottle that contained only water and air represented the normal level of carbon dioxide in the atmosphere and the control for the experiment ("normal"). The temperature in this bottle increased by 40°F, and this serves to demonstrate the greenhouse effect, the energy from the heat lamp being trapped in the plastic bottle. The temperature of the air in the bottle with the lower amount of atmospheric carbon dioxide rose by 33.5°F ("less"). A reduced amount of carbon dioxide, a lower temperature, less greenhouse effect. The third (Alka-Seltzer) bottle contained more atmospheric carbon dioxide ("more"). The temperature increase in this bottle was 50°F. A higher level of carbon dioxide led to a greater temperature increase when compared to the normal, control bottle.

This experiment clearly demonstrates the greenhouse effect and increased temperature resulting from different levels of carbon dioxide in the air inside each bottle. The warming effect of atmospheric carbon dioxide can be shown in this relatively simple yet creative and fairly unsophisticated demonstration. As all good scientists sometimes do, the students designed a series of more elaborate and quantitative experiments that they could perform in the future. If they had conducted these experiments they would have measured the actual amount of carbon dioxide in each of the bottles using special carbon dioxide–detecting instruments. Unfortunately, such instruments are not readily available in a high-school laboratory.

This elegant little experiment was not designed by some crazy, white-haired guy in a lab coat, but by a group of typical high school students who were asked an atypical question as part of a critical thinking challenge. They were acting like scientists because they were being treated like scientists. Maybe one of them might become a scientist and design the ultimate fuel-efficient vehicle or discover the catalyst that could be used in the solar electrolysis of water to generate hydrogen for future energy needs.

You do not need expensive or fancy equipment to perform the students' experiment. Most people can get hold of plastic soda bottles, thermometers, and effervescent tablets. It might be a little harder to get the hydroxide solution, but lime that is used on gardens might work—experiment! Or better still, design a different experiment: perhaps one that uses sunlight as the energy source instead of the heat lamps which use electricity that results in the emission of carbon dioxide! Of course, that electricity could be generated by solar panels rather than by a fossil fueled power plant.

Environmental science student using a systems approach to investigate an inquiry-based curriculum unit on air pollution during a student summer enrichment institute.

What Is My Ecological or Carbon Footprint?

Every one of us uses the earth's natural resources to a greater or lesser extent. In 1995, Mathis Wackernagel and William Rees suggested that we can assess a country's impact on the environment by determining it's *ecological footprint*. For example, the developed nations are responsible for consuming nearly 75 percent of the earth's ecological capacity, in the following percentages:

United States	25%
European Union	19%
China	18%
India	7%
Japan	5%

An individual's ecological footprint depends upon how much he or she consumes to maintain his or her lifestyle. People in many developed countries can be considered to be suffering from "affluenza"; they have adopted lifestyles that require vast amounts of the earth's ecological capacity in terms of energy and resource needs. People in less affluent regions generally use far fewer resources.

Ecological footprint is generally calculated in numbers of hectares of the earth that is required to sustain an individual. For example, a person living in America requires around 10 hectares; in Europe or Japan around five hectares; in China around two hectares, and in India less than one hectare.

So what does this mean? A person living the average American lifestyle needs roughly ten times more than the average person living in India, and twice as much as his or her European and Japanese counterparts! Currently, these estimates mean that we are living beyond our means from an ecologically sustainable viewpoint, and already require more resources than one Earth can provide in the long term. If all the people living in India and China adopted the lifestyle of the average American, we would definitely need several more Earths to provide the necessary resources. As we all know, there is only one planet Earth that is the home for every human, animal, and plant. Simply put, we cannot go on with business as usual.

Each of us has an ecological footprint, and each of us is responsible for a certain amount of carbon dioxide resulting from decisions that we make on a daily basis in our own lives. Most people are unaware of just how much of an impact they have and the volume of carbon dioxide they are contributing from driving their cars or from the electricity they use in their homes.

I strongly encourage you to take a test and estimate your own carbon or ecological footprint by visiting a Website that will calculate your personal score by asking a number of simple questions relating to your lifestyle habits.

The Earthday Network has a calculator that can be found at www.myfootprint.org

The World Resources Institute provides one at www.safeclimate.net/calculator/

The Environmental Protection Agency has a list that contains a wide selection to choose from at *www.epa.gov/climatechange/emissions/individual.html*

You can also get an idea of the volume of carbon dioxide that you emit from driving your car and the electricity usage in your home by the following quick, simple, and fun exercises that help you visualize the amount of carbon dioxide emissions in terms of a certain number of refrigerators full of carbon dioxide that you leave along the side of the road or on your doorstep!

How Many Refrigerators Full of Carbon Dioxide Do I Emit by Driving My Car?

Here are some scientific facts:

One gallon of gasoline produces approximately 20 pounds of carbon dioxide.

One pound of carbon dioxide has a volume of roughly 8.7 cubic feet at standard temperature and pressure.

One household refrigerator has a volume of 18 cubic feet.

From this we can calculate that:

> One pound of carbon dioxide almost fills half a refrigerator (8.7 cubic feet/18 cubic feet), so two pounds of carbon dioxide just about fills an entire fridge

> the 20 pounds of carbon dioxide that one gallon of gasoline produces takes up the space of 10 refrigerators

For each gallon of gas you use in your car, you leave 10 refrigerators full of carbon dioxide along the side of the road.

This can be expanded to the following interesting relationship between miles per gallon and refrigerators of carbon dioxide left along the side of the road for every mile, and for every one hundred miles driven, and for every year. The average American car is driven 12,500 miles a year, which means that for the following fuel efficiencies, the corresponding number of refrigerators full of carbon dioxide would be left at the side of the road.

Miles/gallon	Refrigerators/mile	Refrigerators/100 miles	Refrigerators/year
5	2	200	25,000
10	1	100	12,500
15	0.67	67	8,375
20	0.5	50	6,250
25	0.4	40	5,000
30	0.33	33	4,125
35	0.29	29	3,625
40	0.25	25	3,125
45	0.22	22	2,750
50	0.2	20	2,500

Determining a car's fuel efficiency is easy. Fill up the gas tank and set the trip counter to zero. When you have used up most of the tank, fill it up again, and note how many gallons it took. Note the number of miles that were recorded on the trip counter between fill-ups. Divide the number of miles by the number of gallons of gas and you have the fuel efficiency of your vehicle in miles per gallon. The more you perform this analysis and average the results, the more accurate the fuel efficiency value. Some cars today do the work for you, featuring a fuel efficiency gauge on the dashboard.

For example, if it takes 10 gallons to fill up the tank and the total trip mileage was 200 miles, the car's fuel efficiency is:

200 miles/10 gallons = 20 miles per gallon

If you look at the table on page 124 you can see that for every mile you drive your car you leave half a refrigerator full of carbon dioxide at the side of the road. For every 100 miles driven, you leave 50 refrigerators of carbon dioxide along the highway. If you drive the average number of miles per year, you are responsible for leaving 6,250 refrigerator size containers of carbon dioxide along the road! Of course, if you know someone who has a gas guzzler that gets 10 mpg, he or she leaves twice as many refrigerators along the road, because their vehicle emits double the amount of carbon dioxide (1 refrigerator/mile; 100 refrigerators/100 miles; 12,500 refrigerators a year).

But who's counting?!

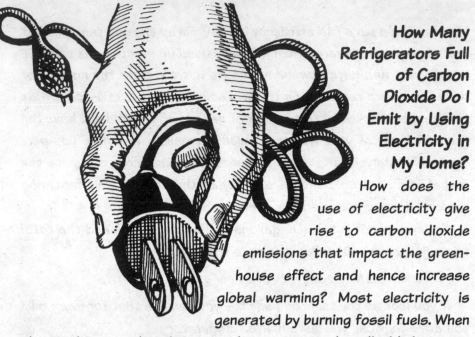

How Many Refrigerators Full of Carbon Dioxide Do I Emit by Using Electricity in My Home?

How does the use of electricity give rise to carbon dioxide emissions that impact the greenhouse effect and hence increase global warming? Most electricity is generated by burning fossil fuels. When they undergo combustion to produce steam, carbon dioxide is among the pollutants released. In 2005, the Energy Information Administration assessed that America's electricity is produced from the following energy sources:

Energy Source	% of total electricity
Coal	49.6
Nuclear	19.3
Natural Gas	18.7
Hydroelectric power	6.5
Petroleum	3.0
Renewable	2.2
Other	0.7

(Let's assume you live in an area that gets its electricity from a coal-burning power plant.)

Here's another scientific fact:

> Producing one kilowatt-hour of electricity from coal pro-
> duces about 2.3 lbs of carbon dioxide.

From this we can calculate that:

> Producing a kWh of electricity results in approximately 20
> cubic feet of carbon dioxide, which for the sake of argument
> is about the average size of one household refrigerator.

For each kilowatt-hour of electricity you use in your home, you leave one refrigerator full of carbon dioxide on your doorstep.

Your monthly electricity bill is the price per kilowatt-hour multiplied by the number of kilowatt-hours used during that month, before any additional costs and taxes are added on.

Check your monthly electric bill to see how many kilowatt-hours you use and how much it costs. Some bills also provide a breakdown of electricity use on a month-by-month basis for the past year. Most of the electricity used in American homes powers air conditioners, lighting, heating, refrigeration, and water heating in that order. A large amount of electricity used in the home comes also from the combined use of small electrical devices, heating elements, and motors, televisions, cooking stoves, clothes dryers, freezers, clothes washers, dishwashers, personal computers, and furnace fans. You can find out if your electricity is generated by a coal- or natural gas-fired plant, a hydro dam, nuclear plant, or wind farm. You could also compare your electricity use to that of your neighbor. What could you do to conserve energy and become more energy efficient and leave fewer refrigerators full of carbon dioxide on your doorstep?

Simple Things to Save Money and the Planet!

Many people could list things that they already know to reduce carbon dioxide emissions. Many people already *do* a lot of things to reduce their own carbon dioxide emissions and help alleviate global warming. If you have read this much of the book, you would probably like to do more. The questions to ask yourself are first, What actions *could* I take? And second, What actions *will* I take?

Remember that no one can do everything. We all contribute carbon dioxide emissions to greater or lesser extents. But each one of us can do something, and no matter how small, little actions can add up and have a large impact. Such a strategy will minimize your own carbon footprint by many refrigerators full of carbon dioxide and help the planet. Some actions have initial costs, but also have cost-saving benefits in the long term. Most people are happy if they can save money on their electricity bill or, with gasoline prices above the $3 a gallon mark, on their pit stops at the pump. The money you save can literally be "saved" for a rainy day, invested in an environmentally conscious stock option. Or you can spend it and stimulate the economy— in environmentally friendly ways of course!

The following list includes just a few of the many, often simple, actions that we can take to save money and save the planet! Check off those that you already knew about!

Saving Money at the Pump

1. Drive less.

2. Walk more.

3. Ride a bicycle.

4. Carpool to work or school.

5. Take turns with friends to drive each other to places like the movies.

6. Combine your trips.

7. Fly less.

8. Take public transport when you can.

9. Drive sensibly.

   Keep tires at the correct pressure.

   Keep the engine tuned.

   Keep the air filter clean.

   Avoid fast starts and sudden stops.

   Use the air conditioner less.

   Fix any leaks in the air-conditioning unit.

   Do not let the car sit with the engine idling.

   Drive with an empty trunk as often as possible.

10. Trade in your old car for a hybrid.

11. Trade in your old car for a more fuel-efficient car.

Saving Electricity in the Home

12. Set the water heater thermostat to 120° F.

13. Put an insulation blanket around the hot-water tank.

14. Install low-flow shower heads and water faucets, then take shorter showers.

15. Wash clothes in warm or cold water and only do full loads, then let the clothes air dry on a clothesline.

16. Only run the dishwasher when it is full and use energy-saving settings, in particular air dry, rather than heat dry.

17. Install a programmable thermostat to control home heating and set it to 68° F during the day.

18. Lower the home heating thermostat at night and use extra blankets (not electric blankets!).

19. Keep the cooling coils on the refrigerator clean and dust free, and check the door seals.

20. Set the refrigerator compartment to 40° F.

21. Set the freezer compartment to between 0–5° F.

22. Make sure the house is well insulated.

23. Use weather stripping around doors and windows.

24. Replace old windows with new, energy-efficient windows.

25. Use the home air conditioner less—open windows and use a fan.

26. Use window shades in the summer and insulated curtains in the winter.

27. Replace old appliances with new Energy Star appliances.

28. Choose gas appliances rather than electric ones.

29. Replace incandescent lightbulbs with compact fluorescent or LED lights, and keep them clean and free of dust (careful disposal of old compact fluorescent bulbs is required).

30. *Do not leave appliances or electric devices plugged in when they are not being used—they still draw "ghost" or "vampire" current!*

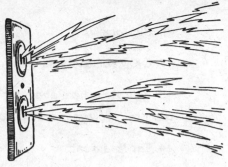

31. Assess the heating system in your home to make sure it is running at its maximum efficiency.

32. Look at using wood stoves or natural gas as a heat source, rather than oil or electricity.

33. Have a home energy audit conducted by your local utility company.

34. If you build a new and smaller home, include all of the latest energy efficiency measures that you can afford, maybe seek LEED certification. New construction methods are being devised on what seems a daily basis—do some research, and find what works best in your area.

35. Monitor your electricity bill and take one step to reduce it each month.

36. Check with your utility company and ask to be supplied with "green" electricity.

Other Helpful Tips to Consider

37. Buy things that are recycled, or better still, that can be reused.

38. Plant a tree, or two, or ten, or a whole forest!

39. Support organizations that are working to reduce global warming and help the environment in general.

40. Pay bills online.

41. Buy locally.

42. Grow some of your own food.

43. Educate others about global warming, and encourage them to follow your example of reducing global warming.

44. Eat less meat.

45. Check the size of your ecological footprint and take steps to reduce it.

46. Decide on where you stand regarding the implementation of any future carbon taxes. Who should pay what, if anything, and why? Keep yourself well informed on the latest news and developments on the global warming issue.

47. Vote your conscience—write, e-mail, and call your local, state, and federal representatives with your opinions.

48. Check out what can be done in your office, workplace, or school to help reduce greenhouse gas emissions and alleviate global warming.

49. Become a wise consumer. Check the labels. Paper or plastic? Neither—take your own bags. Avoid overpackaged items. Reduce, reuse, recycle—remarkable!

50. Don't be a big part of the problem any longer—take action to become part of the solution.

Now that you have checked off all those things on the list that you knew about already, go back and check off all the things that you are doing or will do from now on! Remember, every little negative feedback loop helps global warming in a positive way. Let's all work together to bring about the changes we need, so that our concerns about global warming will become a thing of the past. Has your city signed the Mayors Climate Protection Agreement? Find out! If not, why not? Act *now* before it gets too hot to go back into the kitchen!

An Inconvenient Truth: The Verdict is in!

"In 39 years, I have never written these words in a movie review, but here they are. You owe it to yourself to see this film."
—Roger Ebert, *Chicago Sun-Times*

"One of the most important films of our time."
—Jeffrey Lyons, NBC's *Reel Talk*

"It doesn't matter whether you're a Republican or Democrat, liberal or conservative . . . your mind will be changed in a nanosecond."
—Roger Friedman, *Foxnews.com*

Al Gore has been spreading the word on global warming for many years. His efforts culminated in 2006 with the production of the Academy Award–winning documentary film *An Inconvenient Truth*. The film also won Melissa Etheridge an Oscar for

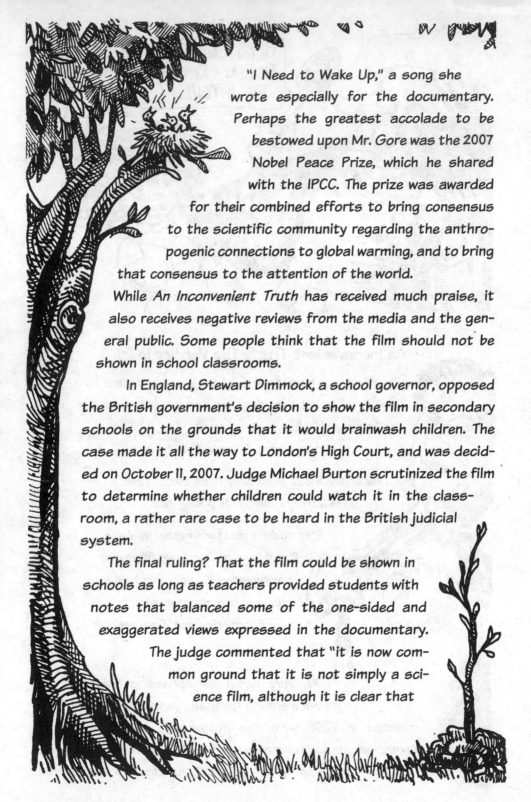

"I Need to Wake Up," a song she wrote especially for the documentary. Perhaps the greatest accolade to be bestowed upon Mr. Gore was the 2007 Nobel Peace Prize, which he shared with the IPCC. The prize was awarded for their combined efforts to bring consensus to the scientific community regarding the anthropogenic connections to global warming, and to bring that consensus to the attention of the world.

While *An Inconvenient Truth* has received much praise, it also receives negative reviews from the media and the general public. Some people think that the film should not be shown in school classrooms.

In England, Stewart Dimmock, a school governor, opposed the British government's decision to show the film in secondary schools on the grounds that it would brainwash children. The case made it all the way to London's High Court, and was decided on October 11, 2007. Judge Michael Burton scrutinized the film to determine whether children could watch it in the classroom, a rather rare case to be heard in the British judicial system.

The final ruling? That the film could be shown in schools as long as teachers provided students with notes that balanced some of the one-sided and exaggerated views expressed in the documentary. The judge commented that "it is now common ground that it is not simply a science film, although it is clear that it is based substantially on scientific research and opinion."

Burton went on to say in his ruling, "It is plainly, as witnessed by the fact that it received an Oscar this year for best documentary film, a powerful, dramatically presented and highly professionally produced film . . . built around the charismatic presence of ex-Vice-President Al Gore, whose crusade it now is to persuade the world of the dangers of climate change caused by global warming." He also pointed out that some of the claims could be viewed as politically partisan and others made in "the context of alarmism."

If you are a teacher in Britain, here are the points that the court ruled need to be addressed in your classroom when you show the film:

> The rate of sea-level rise may not take place as quickly as the film suggests.

> The only drowned polar bears found to date have been four deaths that occurred after a storm.

> Coral bleaching could result from other factors in addition to climate change.

> That it is "very unlikely" that the Atlantic's Gulf Stream will completely shut down.

> So far, nobody has been evacuated from atolls in the Pacific Ocean as a direct consequence of climate change.

> The "exact fit" of the two graphs showing carbon dioxide levels and temperature changes over an extended time period "overstated the case."

> The drying of Lake Chad was "far more likely [the] result [of] . . . population increase, overgrazing, and regional climate change" according to Burton.

> There is no scientific evidence of a direct link between Hurricane Katrina and climate change.

The judge also said in his ruling that "many of the claims made by the film were fully backed up by the weight of science . . . and supported by research published in respected, peer-reviewed journals in accordance with the latest conclusions by the IPCC." He concluded that:

> Climate change is mainly attributed to man-made emissions of carbon dioxide, methane, and nitrous oxide.

> Global temperatures are rising and are likely to continue to rise.

> Climate change will cause serious damage if left unchecked.

> It is entirely possible for governments and individuals to reduce their impact on climate change.

So, the High Court ruled; the verdict is in. Watch the film yourself and be the judge! Or better still, watch it with your children, your friends and relatives, or your students and let them be the jury! I first watched *An Inconvenient Truth* with a group of schoolchildren ranging in age from seven to sixteen years old. At the end, one of the seven-year-olds, who happened to be my son, turned to me and asked, "Daddy, we need to do something about this. What can I do?"

Other Sources of Information

Selected resources for everyone, especially the proponents or believers in global warming!

Books and Magazines

There are many publications relating to the topic of global warming and the environment that can be found on the shelves in book-stores. Here is a list of selected titles for further reading.

Books

Brown, Lester R. PLAN B 3.0: *Mobilizing to Save Civilization*, Washington, DC: Earth Policy Institute, 2008.

Cothron, Julia H., Ronald N. Giese, and Richard J. Rezba. *Students and Research: Practical Strategies for Science Classrooms and Competitions*, 4th ed. Dubuque, IA: Kendall/Hunt, 2005.

Gore, Albert Jr. *An Inconvenient Truth: The Crisis of Global Warming*. Emmaus, PA: Rodale Books, 2007.

Lynas, Mark, *Six Degrees: Our Future on a Hotter Planet*, National Geographic, 2008.

Lynas, Mark, *The Carbon Calculator: Easy Ways to Reduce Your Carbon Footprint*, Collins, 2007.

Mackenzie, Fred T. *Our Changing Planet: An Introduction to Earth System Science and Global Environmental Change*, 3rd ed. Upper Saddle River, NJ: Prentice Hall, 2002.

McKibben, Bill. *The End of Nature: 10th Anniversary Edition*. New York: Random House, 2006.

Miller, G. Tyler Jr. *Living in the Environment: Principles, Connections, and Solutions*, 15th ed. Pacific Grove, CA: Thomson Brooks/Cole, 2007.

Raven, Peter H, and Linda R. Berg, *Environment*, 5th ed. Hoboken, NJ: John Wiley, 2006 de Rothschild, David. *Global Warming Survival Handbook*. Emmaus, PA: Rodale Books, 2007.

Time Magazine editors. *Global Warming: The Causes, the Perils, the Politics—and What It Means for You*, New York: Time Books, 2007.

Wackernagel, Mathis, and William Rees. *Our Ecological Footprint: Reducing Human Impact on the Earth*. Gabriola Island, BC: New Society Publishers, 1995.

Magazines

Two magazines (at least) have devoted issues to global warming; see National Geographic's "Global Warming: Bulletins from a Warmer World" (September 2004) and Backpacker Magazine's "The Global Warming Issue" (September 2007).

Movies

There are a number of films relating to global warming and consumption of the earth's resources.

A Global Warning?, The History Channel, 2007

Affluenza, KCTS/Seattle and Oregon Public Broadcasting, 1997.

An Inconvenient Truth: A Global Warning, Paramount Classics, 2006.

Everything's Cool: A Toxic Comedy About Global Warming!, City Lights Media, 2007.

Six Degrees Could Change the World, National Geographic, 2008.

The 11th Hour, Warner Home Video, 2008.

The Human Footprint, National Geographic, 2008.

"What's Up with the Weather?" NOVA/Frontline, WGBH Boston, 2000.

> Even Hollywood has made global warming a topic for the feature films: *The Day After Tomorrow* (2004, the consequences of climate change), *The American President* (1995, reducing carbon dioxide emissions), and *Waterworld*, (1995, melting of polar ice caps/sea level rise.)

Websites

An Internet search will pop an enormous list of Websites to browse for more on global warming. A few to look at:

www.enviroliteracy.org/ (The Environmental Literacy Council's Website. Click on Air and Climate.)

www.climatecrisis.net (The companion site to An Inconvenient Truth.)

www.epa.gov/ (The official Website of the Environmental Protection Agency. Click on Climate Change.)

www.globalwarmingchallenge.org (The NRDC's Global Warming Challenge Competition: The Difference of One. Encourages kids to make a difference by spreading the word.)

www.ipcc.ch/ (The official Website of the Intergovernmental Panel on Climate Change.)

Selected Resources for Everyone, Especially the Skeptics or Nonbelievers in Global Warming!

For those of you who want to check out the alternative perspective on global warming, here is a book, a movie, and a Website.

Book

The Politically Incorrect Guide to Global Warming and Environmentalism, Christopher C. Horner, Regnery Publishing, 2007.

Movie

The Great Global Warming Swindle, Martin Durkin, WAGtv, 2007.

Website

www.globalwarmingskeptics.info (A Website with the latest anti-climate change perspectives. Edited by James Peden.)

Acknowledgments

Many people have been a source of inspiration at various points in my life. I would like to take the opportunity to thank:

Principal George Crook and chemistry instructor Alan Whitfield at Tamworth College of Further Education, in England, for instilling an inquisitiveness in the sciences that remains with me to this day.

My long-time friend and colleague, Jeannette Adkins of Christchurch School, for her ongoing support and guidance as we probe new arenas of inquiry-based education together.

Headmaster J. E. Beyers and Assistant Head Dr. Neal Keesee, for providing me with the opportunity to make a difference at Christchurch School.

The students in my environmental science class of 2007–2008 at Christchurch School, who rose to the challenge of designing the carbon dioxide experiment referred to in this book.

Tim Knox and Allan Munro for their encouragement and support of my professional development in marine and environmental education, and Joan Michand, Hugh McGraw, and J. C. Boggs, for helping me bring my ideas on inquiry-based education to a wider audience of students and teachers during my time in New Hampshire.

Professor Andrew Friedland, chair of the Environmental Studies Department at Dartmouth College, for enlightening discussions on many aspects of environmental science.

Fred Wetzel, formerly of The College Board, and Beth Nichols and Tom Corley, both of Education Testing Service, for involving me in the Advanced Placement Environmental Science program, and to my colleagues on the Test Development Committee for providing a stimulating, creative, and enjoyable environment in which to work.

Professor Ron Geise, College of William and Mary, for his valuable discussions over the years regarding experimental design strategies.

My wife, Cindy, my son, Taylor, and my daughter, Julia, for giving me a reason to believe in myself.

Finally, a special thanks to David Rogers, who has always been more than a father to me.

About the Author and Illustrator

DEAN GOODWIN, Ph.D. is the executive director of the Center for Inquiry-Based Education. He is also director of marine and environmental education and teaches at Christchurch School in Virginia. Dr. Goodwin is a nationally recognized environmental educator with an honors degree in biochemistry, a post-graduate certificate in science education, and a Ph.D. in mechanistic organic photochemistry. He has more than twenty-five years teaching experience at both the college and secondary level.

JOE LEE is an illustrator, cartoonist, writer and clown. A graduate of Ringling Brothers, Barnum and Bailey's Clown College, he worked for many years as a circus clown. He is also the illustrator for many other For Beginners books including: *Dada and Surrealism For Beginners, Postmodernism For Beginners, Deconstruction For Beginners,* and *The Olympics For Beginners.* Joe lives with his wife, Mary Bess, three cats, and two dogs (Toby and Jack).

THE FOR BEGINNERS® SERIES

AFRICAN HISTORY FOR BEGINNERS: ISBN 978-1-934389-18-8

ANARCHISM FOR BEGINNERS: ISBN 978-1-934389-32-4

ARABS & ISRAEL FOR BEGINNERS: ISBN 978-1-934389-16-4

ASTRONOMY FOR BEGINNERS: ISBN 978-1-934389-25-6

BARACK OBAMA FOR BEGINNERS, AN ESSENTIAL GUIDE: ISBN 978-1-934389-38-6

BLACK HISTORY FOR BEGINNERS: ISBN 978-1-934389-19-5

THE BLACK HOLOCAUST FOR BEGINNERS: ISBN 978-1-934389-03-4

BLACK WOMEN FOR BEGINNERS: ISBN 978-1-934389-20-1

CHOMSKY FOR BEGINNERS: ISBN 978-1-934389-17-1

DADA & SURREALISM FOR BEGINNERS: ISBN 978-1-934389-00-3

DECONSTRUCTION FOR BEGINNERS: ISBN 978-1-934389-26-3

DEMOCRACY FOR BEGINNERS: ISBN 978-1-934389-36-2

DERRIDA FOR BEGINNERS: ISBN 978-1-934389-11-9

EASTERN PHILOSOPHY FOR BEGINNERS: ISBN 978-1-934389-07-2

EXISTENTIALISM FOR BEGINNERS: ISBN 978-1-934389-21-8

FOUCAULT FOR BEGINNERS: ISBN 978-1-934389-12-6

GLOBAL WARMING FOR BEGINNERS: ISBN 978-1-934389-27-0

HEIDEGGER FOR BEGINNERS: ISBN 978-1-934389-13-3

ISLAM FOR BEGINNERS: ISBN 978-1-934389-01-0

KIERKEGAARD FOR BEGINNERS: ISBN 978-1-934389-14-0

LINGUISTICS FOR BEGINNERS: ISBN 978-1-934389-28-7

MALCOLM X FOR BEGINNERS: ISBN 978-1-934389-04-1

NIETZSCHE FOR BEGINNERS: ISBN 978-1-934389-05-8

THE OLYMPICS FOR BEGINNERS: ISBN 978-1-934389-33-1

PHILOSOPHY FOR BEGINNERS: ISBN 978-1-934389-02-7

PLATO FOR BEGINNERS: ISBN 978-1-934389-08-9

POSTMODERNISM FOR BEGINNERS: ISBN 978-1-934389-09-6

SARTRE FOR BEGINNERS: ISBN 978-1-934389-15-7

SHAKESPEARE FOR BEGINNERS: ISBN 978-1-934389-29-4

STRUCTURALISM & POSTRUCTURALISM FOR BEGINNERS: ISBN 978-1-934389-10-2

ZEN FOR BEGINNERS: ISBN 978-1-934389-06-5

www.forbeginnersbooks.com